Cliffs Quick Review

Physical Geology

by
Mark J. Crawford, M.S.

Cliffs®Notes

INCORPORATED

LINCOLN, NEBRASKA 68501

Acknowledgments

Many thanks to Michele Spence, my hands-on editor at Cliffs Notes. In addition to her expert editing, I appreciate her flexibility, patience, and always delicately worded queries.

Cover photograph by John Stuart/The Image Bank

FIRST EDITION

ISBN 0-8220-5335-7

The mysteries of the earth have fascinated the human race for thousands of years. Its restlessness has transformed landscapes, blackened skies, and buried cities. Our way of life is dependent on the geologic resources that we take from mines, quarries, and wells. Expanding our knowledge of our planet, and learning the best ways to use its resources, is critical to the welfare of our ever-expanding population.

The dynamics of the earth are hard to quantify because what can be seen, and physically tested, is limited to the rocks on its surface or taken from drill holes in the crust—a very small part of the planet! The *interpretation* of what these rocks mean—how they relate to each other and how they compare to rocks seen forming today—and the use of new imaging techniques that can explore the subsurface give us our modern view of geology.

More now than ever before our knowledge of geology, geophysics, and other sciences is being directed toward healing the environmental damage that has been done to the earth's surface and groundwater. The field of hydrogeology has grown immensely in the past twenty years in response to our need for clean water supplies. Hydrogeologists explore the complexities of groundwater, how it becomes contaminated, and how to make it clean again.

Today, as in centuries past, the ideas of academic and professional geologists around the world continue to change. Technologic improvements expand the limits of what kind of research can be done. For example, the early theories of continental drift have now been proven, and movements of continental plates can be measured in centimeters.

Despite the incredible advances in knowledge and technology that have made geology one of the most exciting fields, the earth is still startling, surprising, and mysterious. Many more discoveries with worldwide impacts lie ahead.

CONTENTS

CONTENTS

CONTENTS

CONTENTS

CONTENTS

CONTENTS

CONTENTS

CONTENTS

CONTENTS

CONTENTS

CONTENTS

Physical geology is the study of the earth's rocks, minerals, and soils and how they have formed through time. Complex internal processes such as plate tectonics and mountain-building have formed these rocks and brought them to the earth's surface. Earthquakes are the result of the sudden movement of crustal plates, releasing internal energy that becomes destructive at the surface. Internal heat and energy are released also through volcanic eruptions. External processes such as glaciation, running water, weathering, and erosion have formed the landscapes we see today.

Historical Notes

About 2300 years ago, the Greeks, led by the philosopher Aristotle, were among the first to try to understand the earth. During the 1600s and 1700s, scientists believed the earth had been produced by gigantic, sudden, catastrophic events that built mountains, canyons, and oceans.

In the late 1700s, James Hutton, a Scottish doctor, proposed that the physical processes that shape the world today also operated in the geologic past—a principle known as **uniformitarianism.** Another early concept was the **law of superposition**—in an undeformed sequence of sedimentary rocks, each layer is younger than the ones below it and older than those above it. The **law of faunal succession** states that fossils in these rocks occur in the same kind of order, and changes in fossil content represent changes in time. Thus, rocks from different parts of the world containing the same type of fossil formed about the same time. English geologist Charles Lyell enlarged on these ideas and modernized geology with his series of books in the mid to late 1800s.

The Earth's Origin

According to the widely accepted **nebular hypothesis,** the planets and moons in the solar system, including Earth, formed from a huge cloud of mostly hydrogen and helium. Contraction, rotation, and dropping temperatures resulted in the formation of small particles, the first being nickel and iron. These began to stick together, and after tens of millions of years of condensation and accretion, the earth was formed about 5 billion years ago. Although the earth has been cooling ever since and has formed a hard outer crust, part of the interior is still hot and molten.

The Earth's Structure

The earth can be divided into four concentric zones (Figure 1). The innermost is called the **inner core** and is thought to be a solid, spherical mass of iron. Its radius is about 1,216 kilometers (730 miles). The next zone, called the **outer core,** is believed to be a layer of molten liquid rich in nickel and iron that is about 2,270 kilometers (1,362

The Earth's Structure

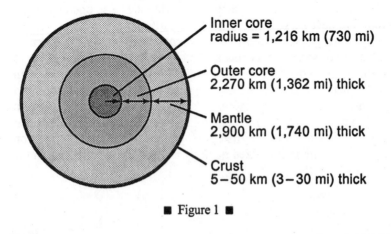

Inner core
radius = 1,216 km (730 mi)

Outer core
2,270 km (1,362 mi) thick

Mantle
2,900 km (1,740 mi) thick

Crust
5 – 50 km (3 – 30 mi) thick

■ Figure 1 ■

miles) thick. The outer core is overlain by the **mantle,** which is a solid yet puttylike rock that can actually flow. The mantle is about 2,900 kilometers (1,740 miles) thick. The **crust,** the outermost zone, is the hardened exterior of the earth and varies in thickness from about 5 to 50 kilometers (3–30 miles).

Continental crust is thicker than **oceanic crust.** The solid **lithosphere** is composed of the crust and the upper part of the mantle. The softer, more flexible part of the mantle underneath the lithosphere is the **asthenosphere** (Figure 2).

The Crust, Lithosphere, and Asthenosphere

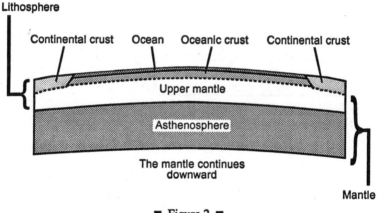

■ Figure 2 ■

As the earth cools, the intense heat being produced in the core creates **convection currents** in the mantle that bring hot mantle material up toward the crust, and colder mantle and crustal rocks sink downward. This heat engine drives **plate tectonics,** or the movements of large segments of the earth's crust (plates) that are separated along deep cracks called **faults.** The plates move over the asthenosphere, which is softer and less resistant. The crust breaks into these segments because of the upward movement of molten material below. The powerful internal tectonic forces squeeze and fold solid rock, creating massive changes in the earth's crust, such as rugged mountains and deep submarine canyons.

The fault boundaries between plates are either convergent, divergent, or transform. A **divergent boundary** is one marked by plates that move away from each other (Figure 3).

A Divergent Boundary

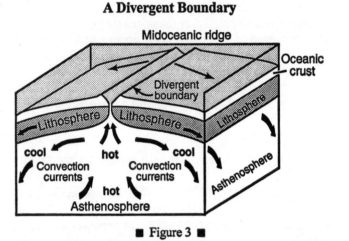

■ Figure 3 ■

A **convergent boundary** is one at which plates come together (Figure 4).

A Convergent Boundary

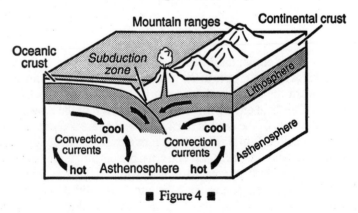

■ Figure 4 ■

Plates slide past each other in opposite directions along a **transform boundary** (Figure 5).

A Transform Boundary

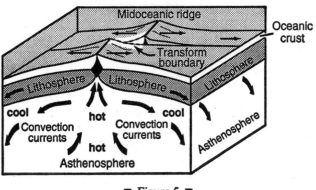

■ Figure 5 ■

New ocean crust is formed along the deep **midoceanic ridges** (divergent boundaries) by the outpouring of mantle lavas on the ocean floor. These ridges are also called **spreading centers.** The new crust pushes to the side the older oceanic crust, which eventually is **subducted,** or forced under another plate at a convergent boundary. The subducted crust moves down a dipping **subduction zone** toward the mantle.

The jostling or rubbing of plates results in high heat flows, volcanic activity, deformation, mountain-building, and earthquakes—creating ideal places to melt rock into magma. Rocks in subduction zones are subjected to friction and higher geothermal gradients that contribute heat to the melting process.

The Earth's Exterior

Various external forces affect the earth's surface, such as different climates and the amount of rainfall. Freezing, thawing, and running water all contribute to **weathering** and **erosion,** processes that break rock down into tiny particles. These particles are then transported by water, ice, or wind as **sediment.** The processes of erosion reduce mountains to hills, create canyons, valleys, and soils, and deposit huge amounts of sediments that either become eroded again or are preserved and lithified into sedimentary rock.

Geologic Time

Geoscientists have estimated the earth to be about **4.5 billion years old.** As the crust cooled, early geologic processes were largely volcanic, building up continental crust and a primitive atmosphere. Bacterial forms of life have been found in rocks that are billions of years old. Complex oceanic organisms such as trilobites began to appear only about 600 million years ago. From about 66 million to 245 million years ago, dinosaurs and other reptiles flourished all over the world. In contrast, human beings have existed in only about the last 3 million years, less than a thousandth of the age of Earth.

The Earth Today

Even though it is nearly 5 billion years old, the earth is still active, and landscapes are constantly changing. The majority of continental rocks have been explored, studied, or sampled. The principle of uniformitarianism—also known as "the present is the key to the past"—is still applicable. Ancient rocks show textures that can be seen forming today from processes such as volcanic eruptions, earthquakes, hot springs, wind, weathering, river action, sedimentation, and erosion.

Technology now allows scientists to probe far deeper than Hutton or Lyell could centuries ago. Underwater cameras, microscopes, geophysical equipment, analytical techniques, sensing devices, and drilling advancements have allowed us to better determine how rocks form. Scientists use uniformitarianism to apply this knowledge to older rocks to better understand the complex history of the earth.

Chemical Composition

The average **chemical composition** of the earth's crust has been
determined from tens of thousands of chemical analyses of rocks and
minerals taken from the surface or drill holes. The most common elements in the crust by weight are oxygen (46.6%), silicon (27.7%),
aluminum (8.1%), iron (5.0%), calcium (3.6%), sodium (2.8%),
potassium (2.6%), and magnesium (2.1%). These eight elements
account for about 98.5 percent of the weight of the crust. The many
other elements from the periodic table make up the remaining 1.5 percent. It may seem surprising that oxygen, which we normally associ-
ate with the atmosphere, is the most abundant element in rocks. It is
an important part of most common minerals, such as quartz (SiO_2)
and calcite ($CaCO_3$).

Minerals and Rocks

Minerals are the building blocks of the earth. A **mineral** is a combination of elements that forms an inorganic, naturally occurring solid
of a definite chemical structure. For example, SiO_2 is always the mineral quartz. A **rock** is a solid material that is composed of various
minerals.

Minerals can have a variety of crystalline shapes. The shape of
the crystal is dependent on the sizes of the atoms of the elements, the
chemical bonds that hold the elements together to form the mineral,
and the pressure and temperature at which the mineral formed.

Most minerals are built around **silica tetrahedrons**—four oxygen atoms connected to a smaller, central silicon atom. Different
arrangements of silica tetrahedrons create distinctive atomic structures in minerals, such as **sheet silicates** (the mica and clay mineral
groups), **chain silicates** (the pyroxene mineral group), or **framework
silicates** (the quartz and feldspar mineral groups).

Only several hundred of the thousands of known minerals are important rock-forming minerals. As one might guess, their chemical compositions contain mostly the eight most common elements in the crust—oxygen, silicon, aluminum, iron, calcium, sodium, potassium, and magnesium. The important rock-forming mineral groups are quartz, feldspars, amphiboles, pyroxenes, clays, micas, and carbonates.

A rock's color is determined by its mineral components: quartz, feldspars, carbonates, and some micas are generally light-colored, tan, or pinkish; pyroxenes, amphiboles, and some micas are dark green to blackish because of their high iron and magnesium content.

Mineral Properties

It is often difficult to identify a mineral simply by looking at it, but each mineral has a set of distinctive characteristics that are easily tested in the field or laboratory.

Hardness. Hardness is a distinctive quality of minerals that is determined by the **Mohs hardness scale.** Talc is the softest mineral on the scale at a value of 1, and diamond is the hardest at a value of 10 (Table 1). Geologists often scratch minerals with a knife blade that has a hardness of about 5. If the mineral scratches the knife, it is harder than 5; if the mineral is scratched, its hardness is less than 5. A thumbnail is about 2.5 on the Mohs scale. Most geologists can remember the hardness scale only by using a mnemonic device (Table 1).

The Mohs Hardness Scale	Mnemonic Device
1. Talc	The
2. Gypsum (thumbnail)	Green
3. Calcite	Clawed
4. Fluorite	Ferocious
5. Apatite (knife blade)	Aardvark
6. Orthoclase	Ordered
7. Quartz	Quick
8. Topaz	Tasty
9. Corundum	Chinese
10. Diamond	Dinners

■ Table 1 ■

Color. Although **color** should always be taken into consideration, minerals can frequently occur in a variety of colors. Chemical weathering also changes a mineral's external color. Scraping a mineral on a porcelain surface, or **streak plate,** leaves a distinctive colored **streak** that is more diagnostic of a mineral than its external color.

Luster. The **luster** is the appearance of the light that is reflected from a mineral's surface. Lusters can be **metallic** (shiny, like gleaming metal), **glassy** or **vitreous** (glazed like porcelain), or **earthy** (dull, not shiny).

Crystal form. Sometimes minerals have a distinctive **crystal form** that reflects a specific internal arrangement of atoms. The crystal form is best developed when the mineral can crystallize slowly from the fluid that contains its elements.

Cleavage. **Cleavage** is the tendency of a mineral to break along preferred crystalline planes that are weakly bonded. The angle between various crystal faces is often distinctive for different mineral groups and can be determined with a magnifying lens in the field.

Other properties. Other properties useful in mineral identification are the way a mineral **fractures,** its **specific gravity** (estimated by how heavy it feels), and whether it is **magnetic** or not.

The Rock Cycle

The **rock cycle** is illustrated in Figure 6. Igneous rocks are produced when molten rock cools and solidifies. When exposed at the earth's surface, the rock is broken down into tiny particles of sediment by weathering and erosion. This weathered material is carried by water or wind to form sedimentary deposits such as beaches, sand bars, or deltas. The sediment is gradually buried by more sediment and subjected to higher pressure and temperature. It eventually hardens into sedimentary rock (**lithifies**). If burial continues, the increasing pressure and temperature at depth recrystallizes the sedimentary rock into a metamorphic rock. The rock cycle is completed when the metamorphic rock becomes so hot that it melts and forms a magma again. Igneous and sedimentary rocks can become metamorphic rocks if they are buried deeply enough or are affected by plate tectonic processes. Metamorphic rocks exposed at the surface will also weather to form sedimentary deposits.

The Rock Cycle

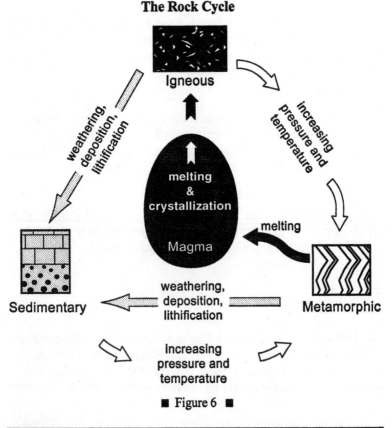

■ Figure 6 ■

The defining characteristic of **igneous rocks** is that at one time they were molten and part of magmas or lavas. A **magma** is a body of molten rock that occurs below the surface of the earth. When magma rises along a deep fault and pours out on the earth's surface, it is termed **lava.** This material then cooled to form a variety of intrusive and extrusive igneous rocks. **Extrusive rocks** crystallized from liquid magmas that reached the surface and were generally vented as volcanic lavas. **Intrusive rocks** crystallized from magmas that did not reach the surface but moved upward into cracks and voids deep in the crust.

Magmatic Differentiation

When a magma cools, chemical reactions occur that create a series of different minerals. This process of **differentiation** occurs along two branches: discontinuous and continuous.

The discontinuous branch. The minerals that form in the **discontinuous branch** are all **ferromagnesian**—that is, they contain high percentages of iron and magnesium, which impart a dark green to black color. The branch is called discontinuous because the minerals form at discrete temperatures and not continuously during cooling. The first mineral to crystallize is **olivine,** followed by **pyroxene, amphibole,** and **biotite.**

The continuous branch. The **continuous branch** is made up of the **plagioclase feldspars.** The calcium/sodium ratio in this mineral type changes continuously as the magma cools. The first feldspars to form contain the highest amounts of calcium; subsequent feldspars have progressively less calcium and more sodium. These minerals tend to be pink, tan, brown, or whitish.

Any magma left over after all these reactions have been completed crystallizes at the lowest temperature as **quartz.**

These theories were first proven in the laboratory by N. L. Bowen in the early 1900s and are also known as **Bowen's reaction series.** The progression in the series explains why the first lavas from a volcanic vent are rich in iron, magnesium, and calcium, are low in quartz, and are dark green to black and why the later lavas are lighter colored and contain more quartz.

Volcanoes and Lavas

Volcanism, or **volcanic activity,** is the venting of liquid magma at the surface of the earth. Occasionally explosive, the process is important in producing continental and oceanic crust. **Volcanoes** are hills or mountains that form around the vent and consist of cooled magma, rock fragments, and dust from the eruptions.

Pieces of rock that are blown out of a volcano are called **pyroclasts** or **pyroclastic debris.** Pyroclasts may also be beads of liquid magma that supercool in the air during descent to form glassy shards of rock. **Pyroclastic flows** are dense, cloudlike mixtures of hot gas and pyroclastic debris that flow down a volcano's sides like an avalanche. These flows can be especially deadly—for example, 30,000 people were killed by a scalding pyroclastic flow on the Caribbean island of Martinique in 1902.

Craters and calderas. The **crater** is the circular depression at the top of the volcano. A **caldera** is a larger depression at least 1 kilometer in diameter that forms at the top of the volcano when the summit is destroyed during an eruption or when the crater floor collapses into the magma chamber below.

Types of volcanoes. There are three kinds of volcanoes: composite, shield, and cinder cone.

Composite volcanoes (stratovolcanoes) have been the sources of some of the more famous and destructive eruptions, such as those of Mount St. Helens, Vesuvius, and Krakatoa. Built up over millions of years, they consist of alternating layers of lava and pyroclastic debris that can approach slopes as steep as 45 degrees. They are characterized by long periods of **dormancy,** or inactivity, that can last for up to hundreds of thousands of years. How violent an eruption is depends on the temperature of the lava and the amounts of silica and dissolved gas in the lava.

Composite volcanoes are located along the **circum-Pacific belt** and the **Mediterranean belt,** which mark the boundaries of colliding crustal plates. The circum-Pacific belt, also known as the "Ring of Fire," runs along the west coasts of South and North America, through the Aleutian Islands south of Alaska, and along the east coasts of Asia and Indonesia.

Shield volcanoes are broad, cone-shaped hills or mountains made from cooled lava flows. The sides are very gently dipping and rarely exceed 10 degrees from the horizontal because the lavas have a **low viscosity** and spread quickly after eruption. (**Viscosity** is defined as resistance to flow; a lava with high viscosity flows sluggishly.) A **spatter cone** is a smaller feature that usually develops on a cooling lava flow from a shield volcano. Gas and lava are ejected through a small vent, building up a steep-sided cone that resembles an appendage.

A **cinder cone (pyroclastic cone)** is composed of pyroclastic material (not lavas) ejected from a vent and commonly has slopes of about 30 degrees.

Volcanic domes. If a magma is thick and viscous and does not easily flow, it may form a **volcanic dome.** Volcanic domes are steep sided or rounded and form near the volcanic vent, creating a plug that can trap gases, build up internal pressures, and lead to violent explosions.

Lava floods. Nonvolcanic lavas called **lava floods** or **plateau basalts** are often associated with deep cracks in the continental crust. Although volcanoes don't form, huge amounts of very nonviscous, "runny" lavas pour from the rift and spread for hundreds of square kilometers. Repeated outpourings of lava have reached thicknesses of 2 kilometers or more in the geologic past.

Extrusive Rock Types

The kind of rock an extrusive lava makes is largely dependent on the chemistry of the venting magma (Figure 7).

Intrusive and Extrusive Rock Classification

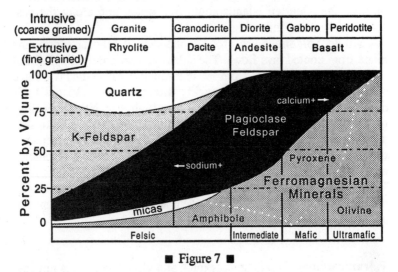

Intrusive (coarse grained)	Granite	Granodiorite	Diorite	Gabbro	Peridotite
Extrusive (fine grained)	Rhyolite	Dacite	Andesite	Basalt	

■ Figure 7 ■

Basalt, andesite, and rhyolite. **Basalt,** usually the first lava to form, contains a high percentage of ferromagnesian minerals and about 25 to 50 percent silica, making it dark green, gray, or black. **Andesite** is a lighter greenish-gray and has more silica and plagioclase feldspar and less ferromagnesian minerals than basalt. **Rhyolite** is the most silicious of the extrusive rocks, containing at least 65 percent silica (mostly in feldspar minerals and quartz) and few ferromagnesian minerals. This chemistry gives it a tan, pink, or cream color. **Dacite** has a composition that falls between those of andesite and rhyolite—it has slightly less potassium feldspar and quartz and slightly more ferromagnesian minerals than rhyolite. Dacite is generally a light grayish-green and often difficult to distinguish from rhyolite in the field.

Mafic, felsic, and intermediate extrusive rocks. More general terms for these rocks are mafic, felsic, and intermediate. **Mafic rocks** have about 50 percent silica and high amounts of iron, magnesium, and calcium and are dark in color. A common mafic rock is basalt. **Felsic rocks** are rich in silica, potassium, sodium, and aluminum and contain only small amounts of iron, magnesium, and calcium. Typical felsic rocks are dacite and rhyolite. Felsic magmas are the most viscous because of their high silica content. **Intermediate rocks,** such as andesite, fall between the mafic and felsic classifications.

Ultramafic extrusive rocks. A less common group are the **ultramafic rocks,** which consist almost entirely of ferromagnesian minerals and have no feldspars or quartz. They contain less than 45 percent silica, and are believed to originate from the mantle. These are some of the least viscous lavas because of their low silica content. A **komatiite** is a typical ultramafic extrusive rock that is mostly olivine and pyroxene, with lesser feldspar.

Rock Textures

The origin of a rock can often be detected from its **texture**—the sizes and orientations of its mineral or rock fragment components. Most extrusive rocks are **fine grained,** meaning their mineral components (grains) are less than 1 millimeter in diameter. Lava flow rocks typically have a **chilled margin** that is very fine grained, or **aphanitic.** Grain size then increases progressively toward the center of the flow. Thicker flows can have medium- to coarse-grained centers.

A **porphyritic** rock contains coarser-grained crystals (**phenocrysts**) that are supported in a **matrix (groundmass)** of finer-grained minerals. The larger minerals had already crystallized and were extruded with the magma, which then rapidly cooled to form the groundmass. **Obsidian (volcanic glass)** is a hard, supercooled, very fine-grained volcanic rock composed of silica.

Basalt flows that have a ropy surface are called **pahoehoe flows** and form when the lava's exterior quickly solidifies into rock. An **aa** (pronounced *ah-ah*) **flow** develops a partially solidified surface as it moves forward. Continued advance breaks the solidified flow's top and sides into a rough, rubbly mass.

Magmas often contain dissolved gas because of higher pressures deep underground. When the magma is suddenly released and vents at the surface, the gas "bubbles" out of the magma, creating numerous holes, cavities, or voids called **vesicles. Pumice** is a volcanic rock that has so much internal void space from gas bubbles that it floats in water. **Scoria** is a very vesicular basalt that contains more gas space than rock and has a very rough, irregular, and pocked exterior.

The lithification of ejected rock fragments and other pyroclastic material creates a variety of **fragmental textures. Dust** and **ash** are the finest-grained particles, followed by **cinders** (pea sized), **lapilli** (walnut sized), and **bombs** or **blocks,** which can be up to a meter across or larger. Blocks are ejected pieces of hardened lavas; bombs are semimolten pieces of lava that solidify as they fall. Small crystals (generally feldspars) that had been formed in the magma before it was ejected are also deposited with the other pyroclastics. A **tuff** is composed of fine-grained pyroclastic material and is named by the most distinctive component, such as an **ash tuff** or **crystal tuff.** A **welded tuff** is a rock that consists of ash particles and glass shards that were hot enough to fuse together when it was deposited. The rocks that contain the larger bombs are called **tuff breccias** or **agglomerates.**

Other distinctive extrusive rock textures occur in flood basalts and submarine lava flows. Flood basalts cool and contract to form vertical, parallel, generally six-sided columns called **columnar structures** or **columnar jointing** (Figure 8). As a submarine lava flow cools, blobs of lava may break through the exterior and harden immediately in the cold water, forming small rounded shapes called **pillow structures.** These are especially useful to the geologist for determining that the rock was formed on the ocean floor and for indicating the base of the flow (Figure 9).

Columnar Jointing

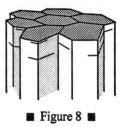

■ Figure 8 ■

Pillow Structures

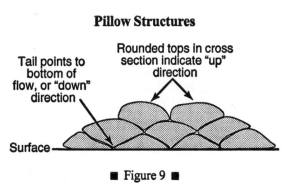

Tail points to bottom of flow, or "down" direction

Rounded tops in cross section indicate "up" direction

Surface

■ Figure 9 ■

Intrusive Rock Types

Intrusive rocks crystallize from magmas that have been intruded into the earth's crust at depths far below the surface. These **intrusions** are then usually exposed millions or billions of years later through the processes of uplift, mountain-building, and erosion. Other intrusive rocks are discovered through deep-drilling programs. **Country rock** is the surrounding rock that the magma invades. A **contact** then separates the cooled intrusive rock from the country rock. Contacts are rarely straight lines, are quite irregular, and mark the change in rock type. The edge of the intrusive rock is usually very fine grained because it is here where the most rapid cooling took place. This edge of the intrusion is called the **chill zone.** The grain size in the intrusion increases away from the chill zone toward the center, where it remained the hottest for the longest time. The intrusive rock often contains **xenoliths**—fragments of the country rock that were torn away during the emplacement of the magma and that are generally most abundant near the contact with the country rock.

Plutonic rocks. Intrusive rocks that were formed deep in the earth's crust are called **plutonic rocks** and are generally coarse grained (mineral grains greater than 1 millimeter in diameter), large, and often associated with mountain-building.

Mafic, felsic, and intermediate intrusive rocks. Intrusive rocks are classified the same way extrusive rocks are—according to the relative amounts of feldspars, quartz, and ferromagnesian minerals (Figure 7). **Gabbro** is a mafic rock and has the same chemistry and mineralogy as basalt; **diorite** is an intermediate rock equivalent to andesite; and **granite** is a felsic rock equivalent to rhyolite. For example, a magma that would form rhyolite if it vented at the surface would crystallize into a granite in a subterranean chamber kilometers below the surface. Granite is the most common intrusive rock on the continents; gabbro is the most common intrusive rock in oceanic crust.

Ultramafic intrusive rocks. Ultramafic intrusions are almost completely composed of ferromagnesian minerals, mostly olivine and pyroxene. They contain less than 45 percent silica and are thought to be derived from the mantle. A typical ultramafic intrusion is called **a peridotite.**

Intrusive Structures

Intrusions are also classified according to size, shape, depth of formation, and geometrical relationship to the country rock. Intrusions that formed at depths of less than 2 kilometers are considered to be shallow intrusions, which tend to be smaller and finer grained than deeper intrusions.

Dikes. A **dike** is an intrusive rock that generally occupies a **discordant,** or **cross-cutting,** crack or fracture that crosses the trend of layering in the country rock. Dikes are called **pegmatites** when they contain very coarse-grained crystals—a single such crystal can range in size from a few centimeters to 10 meters in diameter.

Sills. **Sills** are formed from magmas that entered the country rock parallel to the bedding (layering) and are thus **concordant** with the country rock. Sills can sometimes look like volcanic flows that were interbedded with sedimentary units.

Laccoliths. A **laccolith** resembles a sill but formed between sedimentary layers from a more viscous magma that created a lens-shaped mass that arched the overlying strata upward.

Volcanic necks. A **volcanic neck** is the rock that formed in the vent of a volcano at the end of its eruptive life and remains "standing" after the flanks of the volcano have eroded away.

Plutons. **Plutons** are discordant intrusive rocks that formed at great depths. They tend to be large, coarse grained, and irregular in shape. If the intrusion occupies less than 100 square kilometers (60 square miles) at the earth's surface it is called a **stock;** if it is larger than 100 square kilometers, it is termed a **batholith**. Batholiths are usually composed of granite. They have formed over long periods through the accumulation of smaller magma blobs called **diapirs,** which result from localized melting of the crust; the diapirs then slowly move upward toward the surface and coalesce into a larger mass. Granitic batholiths usually form the cores of mountain complexes and are a result of plate tectonic action.

How Different Magmas Form

Both extrusive and intrusive igneous rocks are derived from magmas. The temperature and pressure conditions in the crust and upper mantle influence the melting temperatures of the minerals in the rocks.

Temperature and pressure increase with depth from the surface and eventually reach a point at which rocks melt. The **geothermal gradient** is the rate at which temperature increases with depth. In the upper crust, the geothermal gradient is about 2.5 degrees centigrade for every 100 meters (330 feet). Geothermal gradients are higher in volcanic regions. **Mantle plumes** are "hot spots" in the crust where mantle material has ascended along deep penetrating cracks in the crust and contributes heat for higher-level melting. Country rock can also be melted from the heat of adjacent intrusions.

Friction is a source of heat in areas where large rock masses are grinding against one another—for example, during mountain-building and plate tectonic activity. Heat is also released through the **radioactive decay** of elements such as uranium, a less important process that only marginally raises the geothermal gradient.

Because of higher pressures, temperatures, changes in density, and gases in solution, magmas tend to rise toward the surface through deep cracks and faults. Being more viscous, felsic magmas rise more slowly than mafic magmas. As magma moves upward it begins to cool, and minerals begin to differentiate.

A very hot magma **assimilates** the country rock it is moving through—that is, the country rock in contact with the magma melts and becomes part of the magma. If a magma assimilates a large amount of country rock, the chemistry of the magma changes. Different extrusive and intrusive rock types form from magmas according to the chemistry of the magma and the differentiation reactions that precipitate the various minerals that make up the igneous rock.

Partial melting is the process by which a portion of the magma that is forming from a melting mass of rock separates and rises as a distinct magma. As a rock is being heated, the first liquid that forms contains a high proportion of the minerals that have lower melting temperatures. A good example is basaltic magma, which is thought to be the result of partial melting in the mantle; the remaining magma in the mantle is then ultramafic in composition. If the entire rock melts, and no magmatic phases escape, the earlier-forming and later-forming liquids mix to form a magma that has the same composition as the original rock.

Igneous Rocks and Plate Tectonics

Igneous rocks form from magmas, and most magmas are associated with plate tectonics. Mafic (basaltic) and ultramafic magmas form along the divergent midoceanic ridges and are major components of new oceanic crust. More felsic magmas, such as andesites and rhyolites, are associated with the edges of continental crust at subduction zones along converging plate boundaries. Whether a magma is intermediate or felsic may depend on the relative amounts of oceanic crust and continental crust in the subduction zone that melt to form the magma. The great abundance of granitic intrusions in continental crust is thought to be related to the partial melting of the lower continental crust.

Intraplate igneous activity occurs in the interior of a single continental plate and is thought to be related to mantle plumes (such as the eruptions at Yellowstone National Park) or flood basalts. Intraplate activity is not associated with moving plate boundaries such as subduction zones.

How Sedimentary Rocks Form

Sedimentary rocks cover about three-fourths of the surface of the continents. There are three kinds of sedimentary rocks: clastic, chemical, and organic. **Clastic sedimentary rocks** form from the consolidation of material such as gravel, sand, or clay (sediment) derived from the weathering and breakdown of rocks. **Chemical sedimentary rocks** result from biological or chemical processes, generally under water, that crystallize minerals that accumulate on the sea floor. **Organic sedimentary rocks,** such as coal, have as their major component accumulations of organic remains from plants or animals that make the rock distinctive.

As the sediments become buried under other sediment layers, pressures and temperatures increase. The sediment hardens into a sedimentary rock, or **lithifies,** after it has gone through the stages of compaction, dewatering, and cementation. During **compaction, the** grains of sediment are packed more tightly together. With increasing pressure some of the water between the sediment particles is squeezed out, **dewatering** the sediment. This process reduces the **pore space,** or open spaces between the grains. At this point, pressure and temperature conditions are such that certain minerals, usually calcite or quartz, fill some or all of the pore spaces and adhere to the sediment fragments, **cementing** them into a sedimentary rock.

A **rock formation** is an occurrence of rock with a set of characteristics that distinguishes it from the rocks above or below it. A formation can then be broken down into smaller rock layers called **members.** A sedimentary **contact** is the boundary surface between two different kinds of rocks and is usually a straight line that represents the original surface where one sediment type was deposited on another.

Clastic Sedimentary Rocks

Clastic sedimentary rocks are classified according to the grain size of the sediment and the kinds of rock fragments that make up the sediment (Table 2). Grain size is largely a function of the distance the particle was transported. In general, the greater the distance traveled, the smaller and more rounded the sediment particles will be. This smoothing of rock fragments during transportation is called **rounding.**

Particle Sizes in Clastic Sedimentary Rocks			
Diameter (mm)	**Particle Type**	**Sediment Name**	**Sedimentary Rock**
More than 256	Boulder	Gravel	Conglomerate or Breccia
64 to 256	Cobble		
4 to 64	Pebble		
2 to 4	Granule		
$\frac{1}{16}$ to 2	Sand	Sand	Sandstone
$\frac{1}{256}$ to $\frac{1}{16}$	Silt	Mud	Siltstone, Shale, or Mudstone
Less than $\frac{1}{256}$	Clay		

■ Table 2 ■

Large, coarse, angular pieces of sediment will be deposited near the source area; well-rounded sand grains will have been transported a considerable distance before being deposited; silt, mud, and clay have been carried even farther. This process is called **sorting.**

Coarse-grained rocks. Sedimentary breccia contains an abundance of coarse, angular fragments of gravel that were deposited very near the source area. A **conglomerate** is formed from coarse-grained, rounded pieces of gravel. **Sandstone** is a medium-grained rock that contains rock particles (mostly quartz) about the size of sand. The grains in a **quartz sandstone** are at least 90 percent quartz. **Graywacke** is a type of "dirty" sandstone that consists of more than 15 percent silt-sized or clay-sized (finer-grained) material that imparts a darker or speckled appearance. If a coarse-grained sandstone consists of over 25 percent feldspar grains it is called an **arkose.**

Finer-grained rocks. The finer-grained clastic sedimentary rocks are called shale, siltstone, and mudstone. **Shale** is a smooth, thinly layered rock that is made up of fine-grained silt and clay particles. Shale is considered a **fissile** rock because it splits very naturally along its layers. A **siltstone** contains mostly silt grains and looks very similar to shale but is not as fissile. **Mudstone,** the finest-grained clastic rock, is not well layered, and contains more clay than does shale or siltstone. Most shales, siltstones, and mudstones are tan, brown, gray, or black.

Chemical Sedimentary Rocks

Limestones. The most common **chemical sedimentary rock** is limestone. Composed mostly of the mineral calcite ($CaCO_3$), **limestones** are usually formed by biochemical processes in shallow seawater. Coral and algae are especially important limestone builders. **Oolitic limestones** form in ocean shallows from the accumulation of **oolites,** sand-sized spheres of chemically precipitated calcite that develop in the tidal zone.

Other variations of limestone result from the deposition and cementation of calcium-rich shells, shell fragments, corals, algae, and the remains of tiny marine organisms. **Coquina** is formed from the cementation of large pieces of broken shells. **Bioclastic** and **skeletal**

limestones are fine- to coarse-grained accumulations of a wider variety of shell fragments and fossils. **Chalk** is a very fine-grained bioclastic limestone composed of accumulations of skeletal debris from tiny marine organisms that drifted down to the ocean floor. All of these "redeposited" limestones could be considered clastic sedimentary rocks, as well as organic sedimentary rocks.

Dolomites. Limestones are frequently converted into **dolomites, or dolostones,** during the early stages of compaction, dewatering, and lithification of the limestone sediment. The process of **dolomitization** involves the removal of calcium from the limestone by magnesium-rich solutions and its replacement in the rock by magnesium. Dolomite's chemical formula is $CaMg(CO_3)_2$.

Cherts. Chert (varieties of which are **flint, agate,** and **jasper**) is a hard, glassy sedimentary rock composed of silica that precipitated from water. Chert nodules, also known as **geodes,** are commonly found in limestones and less so in clastic sedimentary rocks. They form in pockets or voids that might have once been occupied by gas or organic material that has since been removed or decomposed. Cherts can also occur as continuous layers in sedimentary rocks. Chert usually composes at least half of a spectacular layered rock called **iron formation,** which crystallized in shallow seas around the world and is an important source of iron.

Evaporites. Evaporites are rocks that are composed of minerals that precipitated from evaporating seawater or saline lakes. Common evaporites are halite (rock salt), gypsum, borates, potassium salts, and magnesium salts.

Organic Sedimentary Rocks

Organic sedimentary rocks form from the accumulation and lithification of organic debris, such as leaves, roots, and other plant or animal material. Rocks that were once swampy sediments or peat beds contain carbon and are black, soft, and fossiliferous. Rich enough in carbon to burn, **coal** is an organic sedimentary rock that is a widespread and important fuel source. Coquina, bioclastic limestone, and skeletal limestone are also technically organic sedimentary rocks but are usually grouped with the other limestones as being chemically precipitated.

Sedimentary Features

Features that were part of the sediments when they were deposited are often preserved when the sediments become lithified. These features are very useful in reconstructing how the sediment grains were transported, where they came from, the age relationships of different layers, and what the environment was like when the sediments were deposited.

Bedding. **Bedding** is often the most obvious feature of a sedimentary rock and consists of lines called **bedding planes,** which mark the boundaries of different layers of sediment. Most sediments were deposited along a flat surface that was roughly parallel with the depositional surface. An exception is **cross-bedding,** where sediments are carried over an edge or slope by a strong surge of water or wind, forming steeper layers. Cross-bedding tends to occur locally within a larger block of rock, and is overlain and underlain by flat-lying beds. **Herringbone cross-bedding** is a distinctive pattern of alternating cross-bedding directions that is reflective of a rhythmic, high-energy environment, such as a tidal zone.

Graded beds are common when a sediment is being deposited by a slow-moving current. The base of the bed consists of coarser material, which settles to the bottom first. The subsequent beds grade upward through sand and silt, to the finest clay sizes at the top. This pattern is typical in submarine **turbidity flows,** where sediments are dislodged and tumble down an ocean floor slope.

Fossils. Fossils are the remains of plants or animals buried in sediments that were later lithified into rock. They can be extremely useful in determining the depositional environment and the age of the rock. The most obvious fossils are those parts of an organism that have been preserved by being replaced by calcite or silica during lithification. A fossil can also be a **cast** that formed when the organic remains dissolved, leaving an opening, or **mold,** shaped like the organism and later filled with calcite or silica. Other types of fossils include tracks, worm trails, feces, and burrows.

Desiccation cracks and ripple marks. Common structures preserved in sedimentary rocks can be seen forming today along beaches and rivers. **Desiccation cracks,** or **mud cracks,** develop when a muddy sediment is exposed to air and begins to dry out, creating a polygonal pattern of cracks. **Ripple marks** are gentle repeated ridges, usually in sand or silt, that are formed perpendicular to the flow of wind or water.

Sedimentary Environments

Sedimentary rocks give us important information about what the world was like millions of years ago, such as the location of the **source,** or **provenance area,** from which the sediment originated, the kinds of source rocks, and the **paleocurrents** (the direction of flow that deposited the sedimentary grains and how the direction changed with time).

Rock types and structures allow the geologist to determine if the sediments were deposited by glaciers, rivers, lakes, deltas, beaches, sand dunes, wind, lagoons, continental shelf currents, reefs, or deeper ocean waters. High-energy environments such as steep river channels usually deposit coarse arkosic sandstones or conglomerates. Beaches and barrier islands consist of well-rounded quartz sandstone. Lower-energy environments like lake beds, deltas, lagoons, and the deep ocean can be identified by the finer-grained rocks such as shale and siltstone. Limestones usually identify marine reef environments.

An integration of this information over a large region leads to the **reconstruction of the depositional environment**—what the region was like in the geologic past. This three-dimensional reconstruction, over what can be thousands of square kilometers, can be detailed enough to identify such events as flooding and fluctuations in sea level that happened hundreds of millions of years ago.

When rocks are subjected to deep burial, tectonic forces such as folding, and high pressures and temperatures, the textures and mineral compositions begin to change. This process, called **metamorphism,** is the solid-state transformation (no melting) of a rock mass into a rock of generally the same chemistry but with different textures and minerals.

Usually the **metamorphic rock** looks quite different from the original rock, called the **parent rock** or **protolith.** Metamorphic rocks often show contorted patterns of folding that indicate they were soft enough to bend **(plastic deformation).** Folding is achieved by the application of great pressure over long periods. The intensity of the metamorphism increases with increasing temperature and/or pressure, and the highest "grade" of metamorphism approaches partial melting of the rock, almost completing the rock cycle.

Factors Controlling Metamorphism

In most cases, the overall chemistry of the metamorphic rock is very similar to that of the parent rock. A quartz sandstone, for example, will metamorphose into a rock that contains a high percentage of silica. A calcite-rich rock such as limestone can metamorphose only into a calcium-rich metamorphic rock. A quartz sandstone cannot metamorphose into a calcium-rich rock.

Temperature and pressure. **Temperature** and **pressure** are important factors in determining the new minerals that form in a metamorphic rock. Different minerals form under different pressure and temperature conditions. As pressures and temperatures change, a mineral reaches the edge of its **stability field** and breaks down to form new minerals that are stable in the new pressure-temperature field. Higher-temperature minerals tend to be less dense than lower-temperature minerals. The higher temperatures also speed up the chemical reactions that take place during metamorphism.

Water. The amount of water available for metamorphic reactions and the length of time involved are important factors in how quickly and intensely metamorphism proceeds. Metamorphic textures and minerals are most likely formed over 10 to 20 million years or longer.

Geostatic pressure. The **geostatic pressure,** or **confining pressure,** is the pressure that is equally applied to all sides of a deeply buried mass of rock. Geostatic pressure increases with depth.

Differential stress. **Differential stress** is usually the result of tectonic forces applied to a body of rock from different directions. This stress "stretches out" the rock mass into an elongate shape (Figure 10). Generally, the greater the differential stress, the greater the degree of stretching. Components of the rock, such as crystals, fragments, or pillow structures, will also be stretched out, often to the point where they are difficult to recognize.

Differential Stress

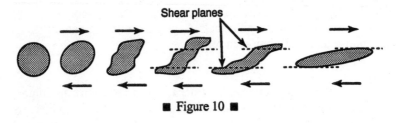

■ Figure 10 ■

Compressive stress. In contrast, a **compressive stress** is applied from directly opposite directions and compresses and flattens the rock mass (Figure 11).

Compressive Stress

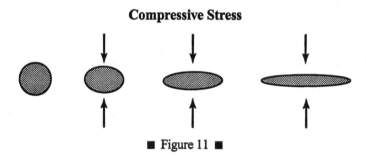

■ Figure 11 ■

Shearing. **Shearing** is related to differential stress and forces parts of the rock mass (usually minerals) to align or grow along a **shear plane.** Shear planes become zones of weakness along which mineral grains are subjected to crushing or recrystallization. Water can enter rocks along shear planes, which speeds up the metamorphic chemical reactions.

Foliation. Prolonged compressive stress and differential stress and/or shearing forces the mineral grains in a metamorphic rock to form parallel layers or bands. This resulting alignment is called **foliation.** New metamorphic minerals crystallize along this foliation. The angle of the foliation is related to the direction of the stress and may cross-cut the original bedding in the rock. A foliation can be so prominent that the original bedding is impossible to identify.

A rock has a **slaty cleavage** if it splits easily along abundant, parallel foliation planes. A **schistose** foliation is more massive and is identified by coarser-grained minerals that have grown along the foliation planes. A **schist** can also be broken along foliation planes, but they are more widely spaced than those in a **slate.** A **gneissic texture** is common in intensely metamorphosed rocks where pressures and temperatures were so high that the rock became plastic, or soft, allowing new minerals to form distinctive light and dark bands.

Types of Metamorphism

There are two major kinds of metamorphism: regional and contact.

Regional metamorphism. Most metamorphic rocks are the result of **regional metamorphism** (also called **dynamothermal metamorphism**). These rocks were typically exposed to tectonic forces and associated high pressures and temperatures. They are usually foliated and deformed and thought to be remnants of ancient mountain ranges.

Metamorphic grades. The different groups of minerals, or **assemblages,** that crystallize and are stable at the different pressure and temperature ranges during regional metamorphism distinguish distinct **metamorphic grades,** or **facies.** The grades are usually named for the dominant minerals or colors that identify them (Figure 12).

In general, proceeding from **low grade** (lower pressure and temperature) to **high grade** (higher pressure and temperature), the following facies are recognized:

- **Zeolite:** low temperature, low pressure

- **Prehnite-pumpellyite:** low temperature, low-medium pressure

- **Greenschist:** low-medium temperature, low-medium pressure

- **Blueschist:** low-medium temperature, high pressure

- **Amphibolite:** medium-high temperature, medium-high pressure

- **Granulite:** high temperature, high pressure

Regional Metamorphic Rock Facies

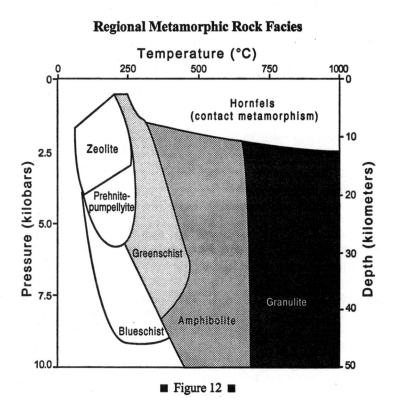

■ Figure 12 ■

Contact metamorphism. Contact metamorphism (also called **thermal metamorphism**) is the process by which the country rock that surrounds a hot magma intrusion is metamorphosed by the high heat flow coming from the intrusion. The zone of metamorphism that surrounds the intrusion is called the **halo** (or **aureole**) and rarely extends more than 100 meters into the country rock. Geostatic pressure is usually a minor factor, since contact metamorphism generally takes place less than 10 kilometers from the surface.

Metamorphic Rock Types

Metamorphic rocks are classified by texture and by mineral composition.

Foliated metamorphic rocks. If a rock is foliated, its name is determined by the type of foliation present and the dominant minerals—for example, a **kyanite schist.** If the minerals are segregated into alternating light-colored and dark-colored layers, the rock is called a **gneiss. Slates** are generally fine-grained, dark-colored, metamorphosed sedimentary rocks that split easily along slaty foliations and were formed under low-grade temperature and pressure conditions. **Phyllites** are slightly more metamorphosed than slates and contain mica crystals that impart a glossy sheen. A **schist** is coarser grained than phyllite or slate and has aligned minerals that can be identified with the naked eye. Some varieties of schist are mica, garnet-mica, biotite, kyanite, and talc schist. A schistose rock composed of the mineral serpentine is called a **serpentinite.**

Migmatites form when temperatures are hot enough to partially melt the rock. The magma is sweated out, or injected, as layers between foliation planes in the rock.

An example of the categories a shale would pass through as temperatures and pressures increase (from low grade to high grade) is as follows: shale/slate/phyllite/mica schist/gneiss/migmatite.

Nonfoliated metamorphic rocks. If a rock is not foliated, its name is derived from its chemical composition. A quartz-rich rock such a sandstone, for example, is called a **quartzite** when it has been metamorphosed. A metamorphosed limestone is called a **marble.** When rocks (especially shales and basalts) are affected by contact metamorphism, they often develop a texture called **hornfels.** A hornfels rock is characterized by evenly distributed, very fine-grained mica crystals that give it a more massive, equigranular appearance.

Hydrothermal Rocks

Hydrothermal essentially means "hot water." **Hydrothermal rocks** are those rocks whose minerals crystallized from hot water or whose minerals have been altered by hot water passing through them. Thus, these rocks are distinct from metamorphic rocks, which are created by solid-state mineral transformations. In fact, many hydrothermal rocks (such as those that form from hot springs and geysers or crystallize as veins in cracks in other rocks) actually build up in layers, much as sedimentary rocks do.

Veins result when hot water moves through cracks in the bedrock of the crust. The water leaches elements from the rocks it passes through. Various minerals are precipitated on the sides of the crack as the temperatures decrease. The shape and orientation of the minerals depends on the temperature, pressure, and rate of flow. When all the available space in the crack has been filled with mineral deposits, the crack is sealed and the vein is complete.

The water involved in hydrothermal processes is usually either seawater that is moving downward through oceanic crust near mid-oceanic ridges or meteoric water. **Meteoric water** is water that is derived from the atmosphere as rain or snow and that moves down into the bedrock from the earth's surface. Water trapped in the original sediments during deposition and lithification (**connate water**) can also be included in hydrothermal reactions but is not a major source of hydrothermal fluid. **Magmatic water** derived from magmas is also a minor component.

The water is heated to very high temperatures as it moves deeper into the crust. It eventually rises again, often removing elements from the rocks it passes through and carrying them in solution. As the hot water rises toward the surface, it begins to cool. This temperature drop induces a number of chemical reactions, and hydrothermal minerals are precipitated.

Metasomatism is the process by which hot-water solutions carrying ions from an outside source move through a rock mass via fractures or pore space. Some of the rock mass is usually dissolved away, and the ions introduced by the water are incorporated into the new minerals that precipitate. Unlike metamorphism, metasomatism can significantly change the overall chemistry of the parent rock. Elements commonly added during metasomatism are iron, sodium, potassium, oxygen, and silica. Easily soluble elements, such as calcium and magnesium from limestones, are often dissolved and carried away, creating more room for new chemical reactions.

Metamorphism and Plate Tectonics

Metamorphic rocks result from the forces active during plate tectonic processes. The collision of plates, subduction, and the sliding of plates along transform faults create differential stress, friction, shearing, compressive stress, folding, faulting, and increased heat flow. The tectonic forces deform and break the rock, creating openings, cracks, faults, breccias, and zones of weakness along which magmas can rise. Generally speaking, the greater the tectonic forces, the higher the pressures and temperatures affecting a rock mass and the greater the amount of resulting structural deformation and metamorphism.

Geologic structures are usually the result of the powerful tectonic forces that occur within the earth. These forces fold and break rocks, form deep faults, and build mountains. Repeated applications of force—the folding of already folded rocks or the faulting and offsetting of already faulted rocks—can create a very complex geologic picture that is difficult to interpret. Most of these forces are related to plate tectonic activity. Some of the natural resources we depend on, such as metallic ores and petroleum, often form along or near geologic structures. Thus, understanding the origin of these structures is critical to discovering more reserves of our nonrenewable resources.

Structural geology is the study of the processes that result in the formation of geologic structures and how these structures affect rocks. Structural geology deals with a variety of structural features that can range in size from microscopic (such as traces of earlier folds after multiple events of deformation have occurred) to large enough to span the globe (such as midoceanic ridges).

Tectonic Forces

Rocks are under **stress** when they are subjected to a force at depth. When the rocks are exposed at the surface after uplift and erosion, the effects of the stress can be studied. Stressed rocks show varying degrees of **strain**—the change in the volume and/or shape of the rock because of that stress. For example, a volcanic agglomerate may be compacted and its pyroclastic fragments stretched (strained) in response to a tectonic stress, such as compression.

Stresses. Three kinds of stress can be applied to rocks: tensional, compressive, and shear. **Tensional stress** occurs when a rock is subjected to forces that tend to elongate it or pull it apart; a rock that has experienced tensional stress tends to be narrower and longer than its original shape, like a piece of gum or taffy that has been pulled. A **compressive stress** on a rock is applied from opposite sides and has a tendency to shorten (compress) the rock between the opposing stresses, which may also stretch it parallel to the stress-free direction. A **shear stress** results when forces from opposite directions create a shear plane in an area in which the forces run parallel to one another. (See Figure 10, p. 32.) The scale of shear stress can vary from a few centimeters to hundreds of meters.

Strains. When subjected to stress, a rock can undergo one of three kinds of deformation (strain): elastic, brittle, or plastic. Deformation is called **elastic strain** if the body of rock returns to its previous shape after the stress has been removed. A good example is the slow rebound of the North American crust after having been downwarped by the great weight of the Pleistocene glaciers. **Brittle strain** occurs when the stress is great enough to break (fracture) the rock. **Plastic strain** results in a permanent change in the shape of the rock. A **ductile** rock is one that "flows plastically" in response to stress. Whether the strain is plastic or brittle depends on both the magnitude of the stress and how quickly the stress is applied. A great stress that is slowly applied often folds rocks into tight, convoluted patterns without breaking them.

Interpreting Structures

Understanding the formation of geologic structures in a region is important in reconstructing its geologic history. Generally, the greater the number of structures, the more complex the geologic history. Structures often offset, rearrange, or bury blocks of bedrock, making geologic interpretation more difficult. Understanding geologic structures is important not only to those in academic fields, but to those in industrial and engineering fields as well. Knowing the structural history of an area is important for finding petroleum and metallic ore bodies and for determining rock stability (for example, in order to locate dams and nuclear reactors on structurally stable bedrock).

Structural events are often inferred from how the bedrock has moved. For example, the law of original horizontality suggests that sedimentary rocks were deposited as flat-lying layers on a relatively horizontal surface. If these rocks are now exposed at the surface and are still flat-lying, it can be concluded that they were uplifted by an evenly applied tectonic force. If they are tilted in one direction, it can be concluded that the uplifting stress was greater at one end and pushed the layers into an inclined position. Occasionally, although the bedrock is horizontally layered, the sedimentary structures and age relationships shown by fossils indicate the *top layer is the oldest*—this arrangement indicates that somehow the entire sequence has been overturned by tectonic forces and what was the oldest layer on the bottom is now on top.

Mapping in the Field

The ease with which structural geology can be understood is largely dependent on how much of the bedrock is available for study. In areas such as northern Canada, where much of the bedrock has been exposed by glaciation, as much as 75 percent of the bedrock can be walked on and studied. Alternatively, in the southeastern United States, often less than 10 percent of the bedrock is exposed because of abundant weathering, soil cover, and vegetation. Reconstructing the geologic history of an area can be especially challenging (and creative) if little rock is exposed.

Geologists try to find all the bedrock exposures, or **outcrops,** in an area to construct a geologic map. They identify rock types, relationships, textures, features (such as cross-bedding), and structures (such as folds and faults) as well as cross-cutting relationships of intrusive rocks, rock mineral contents, and fossils. Detailed directional measurements along structures, when plotted on a map, can reveal a bigger picture of how the rocks have been folded and faulted.

One of the most useful measurements is the strike and dip of a tilted rock unit (Figure 13). The **strike** of the unit is the direction (compass bearing) of the line formed by the intersection of the tilted bedding plane with the horizontal plane. The **dip angle** is the angle between the horizontal plane and the tilted bedding plane. Compasses equipped with a device called an **inclinometer** can determine the dip angle. The direction of dip is always perpendicular to the strike direction. For example, in Figure 13 the rock strikes north-south and dips 25 degrees to the east. A rock that is perfectly flat-lying has no strike direction and no dip. A rock unit that has been tilted into a vertical position has a maximum dip of 90 degrees.

Strike and Dip

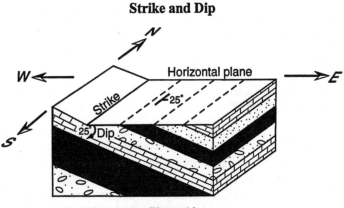

■ Figure 13 ■

A **plan** (two-dimensional) **geologic map** shows the locations and shapes of the outcrops at an appropriate scale and indicates, through a variety of geologic symbols, features such as folds, faults, contacts between different rock units, and strike and dip. A **geologic cross section,** a vertical slice across the map area, can be constructed from the structural information on a geologic map. It depicts the spatial relationships of the rock units and structures beneath the surface (Figure 14). A cross section supplies a third dimension to the plan geologic map. A good geologic map is critical to understanding and interpreting

structures, when they formed, and how they fit into the overall geologic picture.

A Vertical Cross Section

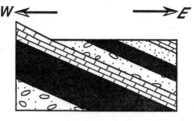

■ Figure 14 ■

Folding

A layered rock that exhibits bends is said to be **folded.** The layered rock was at one time uniformly straight but was stressed to develop a series of arches and troughs. A compressive stress compacts horizontal rock layers and forces them to bend vertically, forming fold patterns.

Anticlines and synclines. An **anticline** is a fold that is arched upward to form a ridge; a **syncline** is a fold that arches downward to form a trough (Figure 15). Anticlines and synclines are usually made up of many rock units that are folded in the same pattern. The tip of a fold is called the **nose.** The center axis of a fold is called the **hinge line** and lies in the **axial plane** that separates the rocks on one side of the fold from the rocks on the other side that dip in the opposite direction. Extensive folding is represented by a repeated pattern of anticlines and synclines. Two anticlines are always separated by a syncline, and two synclines are always separated by an anticline. One side of the fold is called the **limb;** a side-by-side syncline and anticline share a limb. Frequently, an anticline or syncline can be identified only from the systematic change in the dips of the sloping rock units from one direction to the other, identifying the hinge line of the fold (Figure 15).

An Anticline and a Syncline

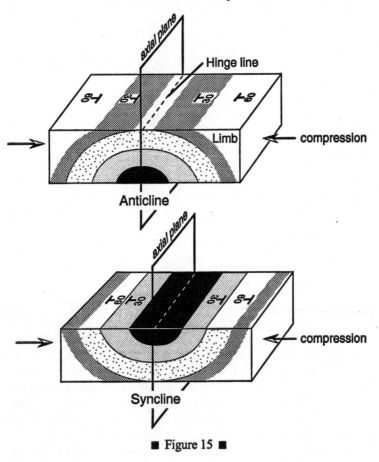

■ Figure 15 ■

Plunging folds. **Plunging folds** have been tipped by tectonic forces and have a hinge line not horizontal in the axial plane. The angle between the horizontal and the hinge line is called the **plunge** and, like dip, varies from less than 1 degree to 90 degrees. Plunging folds characteristically show a series of V patterns on a bedrock surface (Figure 16).

A Plunging Anticline

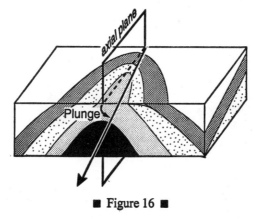

■ Figure 16 ■

Structural domes and basins. A **structural dome,** a variety of anticline, is a feature in which the central area has been warped and uplifted and all the surrounding rock units dip away from the center. Similarly, a **structural basin** is a variation of syncline in which all the beds dip inward toward the center of the basin. Basins and domes can be as large as 100 kilometers across.

Open, isoclinal, overturned, and recumbent folds. A variety of kinds of folds generally reflects increasing amounts of tectonic stress (Figure 17). An **open fold** is a broad feature in which the limbs dip at a gentle angle away from the crest of the fold. **Isoclinal folds** have undergone greater stress that has compressed the limbs of the folds tightly together. The limbs of **overturned folds** dip in the same direction, indicating that the upper part of the fold has overridden the lower part. Depending on where the exposure is in an overturned fold, the oldest strata might actually be on top of the sequence and be misinterpreted as the youngest rock unit. **Recumbent folds,** found in areas of the greatest tectonic stress, are folds that are so overturned that the limbs are essentially horizontal and parallel.

Folds

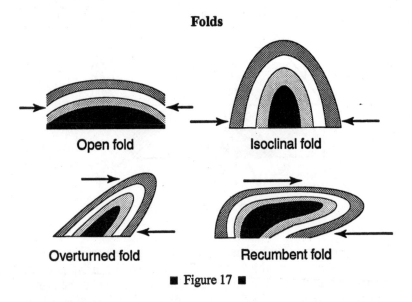

Open fold

Isoclinal fold

Overturned fold

Recumbent fold

■ Figure 17 ■

Fracturing

A rock **fractures** if it is hard and brittle and subjected to sudden strain that overcomes its internal crystalline bonds. If the rock has been displaced along a fracture, such as having one side that is moved up or down, the fracture is called a **fault,** and if there is no displacement along the crack, the fracture is called a **joint.**

Faults. Horizontal or vertical displacement along the fault plane can range from a few centimeters to hundreds of kilometers. The fault can be merely a crack between the two sides of rock, or it can be a **fault zone** hundreds of meters wide that consists of rock that has been very fractured, brecciated, and pulverized from repeated grinding movements along the fault plane. The broken material within a fault is called **fault gouge.** The rocks within a fault zone may also be hydrothermally altered or veined from hot solutions that have migrated up the fault zone. A fault is generally considered active if movement has occurred along it during the past 10,000 years.

Fault movements. Three kinds of fault movements are recognized: dip-slip, strike-slip, and oblique-slip. Movement in a **dip-slip fault** is parallel to the dip of the fault plane in an "up" or "down" direction between the two blocks. The block that underlies an inclined dip-slip fault is called the **footwall;** the block that rests on top of the inclined fault plane is called the **hanging wall.** A **normal dip-slip fault,** or **normal fault,** is one in which the hanging wall block has slipped down the fault plane relative to the footwall block. A **reverse dip-slip fault** is just the opposite: the hanging wall block has moved upward relative to the footwall block (Figure 18).

Dip-Slip Faults

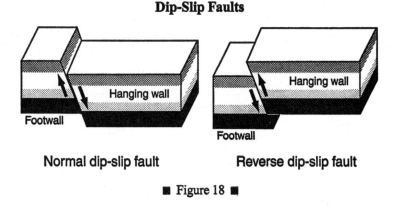

Normal dip-slip fault   Reverse dip-slip fault

■ Figure 18 ■

The blocks on either side of a **strike-slip fault** move horizontally in relation to each other, parallel to the strike of the fault. If a person is standing at the fault and looks across to see that a feature has been displaced to the left, it is called a **left-lateral strike-slip fault**. A **right-lateral strike-slip fault** is one in which the displacement appears to the right when looking across the fault (Figure 19).

Strike-Slip Faults

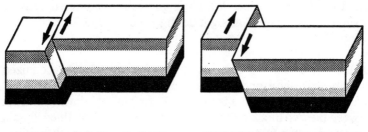

Left-lateral strike-slip fault   Right-lateral strike-slip fault

■ Figure 19 ■

If the fault blocks show both horizontal and vertical displacement, the fault is termed an **oblique-slip**.

A **graben** is formed when a block that is bounded by normal faults slips downward, usually because of a tensional force, creating a valleylike depression. A **horst** results when a block that is bounded by normal faults experiences a compressive force that forces the block upward, forming mountainous terrain (Figure 20).

A Graben and Horst

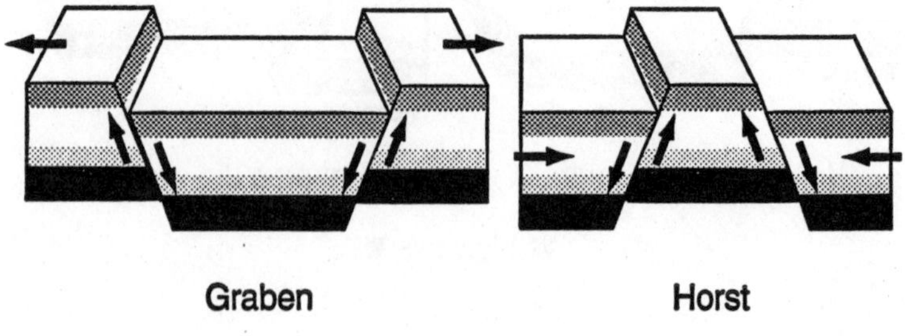

Graben   Horst

■ Figure 20 ■

Thrust faults are reverse dip-slip faults in which the hanging wall block has overridden the footwall block at a very shallow angle for tens of kilometers. The hanging wall block and footwall block of a thrust fault are typically called the **upper plate** and **lower plate**, respectively (Figure 21).

A Thrust Fault

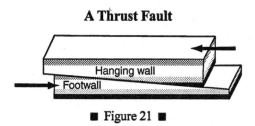

■ Figure 21 ■

Joints. Joints are generally the result of a rock mass adjusting to compressive or tensional stress or cooling. A **joint set** is composed of a series of roughly parallel joints that occur in one direction. Tensional stress usually results in a single joint orientation that is perpendicular to the direction of stress. Compressive stress often generates two cross-cutting joint sets.

Unconformities

An **unconformity** is a contact between two rock units in which the upper unit is usually much younger than the lower unit. Unconformities are typically buried erosional surfaces that can represent a break in the geologic record of hundreds of millions of years or more. For example, the contact between a 400-million-year-old sandstone that was deposited by a rising sea on a weathered bedrock surface that is 600 million years old is an unconformity that represents a **time hiatus** of 200 million years. The sediment and/or rock that was deposited directly on the bedrock during that 200-million-year span was eroded away, leaving the "basement" surface exposed. There are three kinds of unconformities: disconformities, nonconformities, and angular unconformities.

Disconformities. **Disconformities** (Figure 22) are usually erosional contacts that are parallel to the bedding planes of the upper and lower rock units. Since disconformities are hard to recognize in a layered sedimentary rock sequence, they are often discovered when the fossils in the upper and lower rock units are studied. A gap in the fossil record indicates a gap in the depositional record, and the length of time the disconformity represents can be calculated. Disconformities are usually a result of erosion but can occasionally represent periods of nondeposition.

A Disconformity

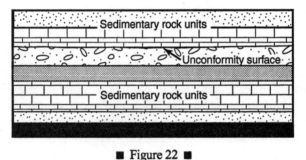

Sedimentary rock units

Unconformity surface

Sedimentary rock units

■ Figure 22 ■

Nonconformities. A **nonconformity** (Figure 23) is the contact that separates a younger sedimentary rock unit from an igneous intrusive rock or metamorphic rock unit. A nonconformity suggests that a period of long-term uplift, weathering, and erosion occurred to expose the older, deeper rock at the surface before it was finally buried by the younger rocks above it. A nonconformity is the old erosional surface on the underlying rock.

A Nonconformity

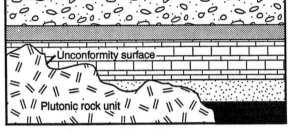

■ Figure 23 ■

Angular unconformities. An **angular unconformity** (Figure 24) is the contact that separates a younger, gently dipping rock unit from older underlying rocks that are tilted or deformed layered rock. The contact is more obvious than a disconformity because the rock units are not parallel and at first appear cross-cutting. Angular unconformities generally represent a longer time hiatus than do disconformities because the underlying rock had usually been metamorphosed, uplifted, and eroded before the upper rock unit was deposited.

An Angular Unconformity

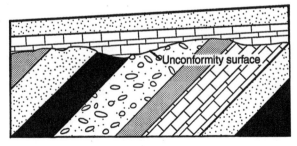

■ Figure 24 ■

Weathering

The process of **weathering** alters rocks at the earth's surface and breaks them down over time into fine-grained particles of sediment and soil. Weathering is the result of the interactions of air, water, and temperature on exposed rock surfaces and prepares the rock for erosion. **Erosion** is the movement of the particles by ice, wind, or water. The particles are then **transported** by that agent until they are **deposited** to form sedimentary deposits, which can be later eroded again or transformed into sedimentary rocks. The weathering of the sediment grains continues during erosion and transportation. Weathering is generally a long, slow process that is continuously active at the earth's surface.

There are two kinds of weathering: mechanical and chemical. **Mechanical weathering** is the process by which rocks are broken down into smaller pieces by external conditions, such as the freezing of water in cracks in the rock. The rock is **chemically weathered** when it reacts with rain, water, and the atmosphere to destroy chemical and mineralogical bonds and form new minerals. For example, feldspar crystals in a volcanic tuff commonly weather to form new clay minerals. Other minerals, such as calcite, dissolve. Quartz, on the other hand, is very resistant to weathering.

Chemical weathering weakens the bonds in rocks and makes them more vulnerable to decomposition and erosion. Thus, the weathered surface of a rock, whether it is an outcrop in the field or the stone walls of a building on a downtown street, looks different from the "fresh" interior of the rock. The most noticeable effect is the discoloration of the surface that results from the breakdown of minerals.

A **spherically weathered boulder** forms when the corners of an angular rock are broken down more quickly than the flat surfaces, forming rounded shapes. **Differential weathering** results when some rocks resist weathering more than other rocks, creating uneven rates of weathering and erosion. This phenomenon often forms arched natural bridges or spectacular mushroom-shaped rock formations, where a broad, more-resistant sandstone ledge is perched on a narrower column of less-resistant shale that has eroded more quickly.

Processes of Mechanical Weathering

Ice. The formation of **ice** in the myriad of tiny cracks and joints in a rock's surface slowly pries it apart over thousands of years. **Frost wedging** results when the formation of ice widens and deepens the cracks, breaking off pieces and slabs. Frost wedging is most effective in those climates that have many cycles of freezing and thawing. **Frost heaving** is the process by which rocks are lifted vertically from soil by the formation of ice. Water freezes first under rock fragments and boulders in the soil; the repeated freezing and thawing of ice gradually pushes the rocks to the surface.

Exfoliation. If a large intrusion is brought to the surface through tectonic uplift and the erosion of overlying rocks, the confining pressure above the intrusion has been released, but the pressure underneath is still being exerted, forcing the rock to expand. This process is called **unloading.** Because the outer layers expand the most, cracks, or **sheet joints,** develop that parallel the curved outer surface of the rock. Sheet joints become surfaces along which curved pieces of rock break loose, exposing a new surface. This process is called **exfoliation;** large rounded landforms (usually intrusive rocks) that result from this process are called **exfoliation domes.** Examples of exfoliation domes are Stone Mountain, Georgia, and Half Dome in Yosemite National Park.

Friction and impact. Rocks are also broken up by **friction** and repeated **impact** with other rock fragments during transportation. For example, a rock fragment carried along in a river's current continuously bounces against other fragments and the river bottom and even-

tually is broken into smaller pieces. This process occurs also during transportation by wind and glacial ice.

Other processes. Less important agents of mechanical weathering include the **burrowing of animals, plant roots** that grow in surface cracks, and the digestion of certain minerals, such as metal sulfides, by **bacteria.** Daily temperature changes, especially in those regions where temperatures can vary by 30 degrees centigrade, result in the expansion and contraction of minerals, which weaken rocks. Extreme **temperature changes,** such as those produced by forest fires, can force rocks to shatter.

Processes of Chemical Weathering

When a rock is brought to the surface millions or billions of years after it has formed, the original minerals that were crystallized deep in the crust under high pressures and temperatures are **unstable** in the surface environment and eventually break down. The primary agents in chemical weathering are water, oxygen, and acids. These react with surface rocks to form new minerals that are stable in, or in **equilibrium** with, the physical and chemical conditions present at the earth's surface. Any excess ions left over from the chemical reactions are carried away in the acidic water. For example, feldspar minerals will weather to clay minerals, releasing silica, potassium, hydrogen, sodium, and calcium. These elements remain in solution and are commonly found in surface water and groundwater. Newly deposited sediments are often cemented by calcite or quartz that is precipitated between the sediment grains from calcium- and silica-bearing water, respectively.

How quickly chemical weathering breaks a rock down is directly proportional to the area of rock surface exposed. Thus, it is also related to mechanical weathering, which creates more exposed surface area by breaking the rock down into pieces, and those pieces into

smaller pieces. The greater the number of pieces, the greater the total amount of surface area exposed to chemical weathering.

Water. Chemical weathering is most intense in areas that have abundant **water.** Different minerals weather at different rates that are climate dependent. Ferromagnesian minerals break down quickly, whereas quartz is very resistant to weathering. In tropical climates, where rocks are intensely weathered to form soils, quartz grains are typically the only component of the rock that remains unchanged. Alternatively, in dry desert climates, minerals normally susceptible to weathering in wet environments (such as calcite) can be much more resistant.

Acids. **Acids** are chemical compounds that decompose in water to release hydrogen atoms. Hydrogen atoms frequently substitute for other elements in mineral structures, breaking them down to form new minerals that contain the hydrogen atoms. The most abundant natural acid is **carbonic acid,** a weak acid that consists of dissolved carbon dioxide in water. Rainwater usually contains some dissolved carbon dioxide and is slightly acidic. The burning of coal, oil, and gasoline releases carbon dioxide, nitrogen, and sulfur into the atmosphere, which react with rainwater to form much stronger **carbonic, nitric,** and **sulfuric acids** that damage the environment (**acid rain**).

Other acids that can affect the formation of minerals in the near-surface weathering environment are **organic acids** derived from plant and humus material. Strong acids that are naturally occurring in the environment are rare—they include the **sulfuric acids** and **hydrofluoric acids** released during volcanic and hot spring activity.

Solution weathering is the process by which certain minerals are dissolved by acidic solutions. For example, calcite in limestone is dissolved easily by carbonic acid. Rain that percolates through cracks and fissures in limestone beds dissolves calcite, making wider cracks that can ultimately develop into cave systems.

Oxygen. Oxygen is present in air and water and is an important part of many chemical reactions. One of the more common and visible chemical weathering reactions is the combination of iron and oxygen to form **iron oxide (rust)**. Oxygen reacts with iron-bearing minerals to form the mineral **hematite (Fe_2O_3)**, which weathers a rusty brown. If water is included in the reaction, the resultant mineral is called **limonite ($Fe_2O_3 \cdot nH_2O$)**, which is yellow-brown. These minerals often stain rock surfaces a reddish-brown to yellowish color.

Soil

The layers of weathered particles of earth material that contain organic matter and can support vegetation are defined as **soil**. Soil can be all or just part of the sedimentary material that covers the bedrock. For example, in Wisconsin the bedrock is covered by up to 120 meters (400 feet) of glacial gravel, sand, silt, and clay—yet only the upper few feet is considered to be soil. It can take hundreds of thousands of years to form soil. Soil contains mostly quartz and clay particles of varying sizes. The quartz sand grains help keep the soil porous, and the clay particles hold water and nutrients for plant growth. **Loam** contains sand, silt, and clay with abundant organic material. **Topsoil** is the upper part of a section of loam, has the highest organic content, and is considered to be the most fertile layer. The rocky **subsoil** underlies the topsoil and contains less organic material.

Residual soil and transported soil. When soil is developed from the weathering of the underlying bedrock it is called **residual soil. Transported soil** is deposited by agents such as ice and water and is not derived from the underlying bedrock. Examples include sand left by retreating glaciers and the mud that is left after a flood. Much of the rich black soil in the midwestern United States was deposited by melting glaciers about 10,000 years ago.

Soil horizons. A fully developed soil, or **mature soil,** consists of three layers, or **soil horizons** (Figure 25). The uppermost layer is called the **A horizon.** Meteoric water moves down through the horizon and typically leaches clay minerals, iron, and calcite from the soil. The top part of the A horizon consists of the dark, organic loamy topsoil (the thin layer of humus and plant litter on the surface that is usually considered part of the A horizon is sometimes referred to as the **O horizon**); the rest of the horizon is a pale yellowish-tan.

The middle layer is called the **B horizon** or the **zone of accumulation.** The leached materials from the A horizon often precipitate in this more clay-rich zone that is reddish-brown from iron minerals. This horizon is also called the **subsoil.** The combined O-A-B section is sometimes called the **solum.** Known as the "true soil," the solum is where most of the biological activity in soil is confined.

Soil Horizons

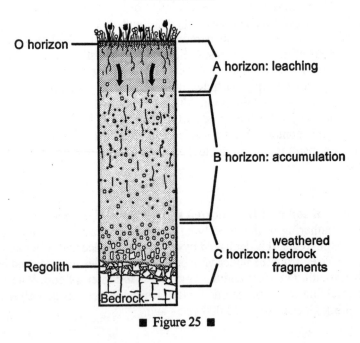

■ Figure 25 ■

The **C horizon** is the lowest soil layer and lies directly above the bedrock. This layer is part soil and part decomposing bedrock fragments. It contains very little organic material. The interface between the bedrock surface and overlying sedimentary material is called the **regolith** and consists of broken, rubbly pieces of bedrock that are variably weathered and decomposed.

Immature soils do not have the complete O-A-B-C profile and are commonly found on steep slopes, where the weathering products migrate down the slope instead of accumulating in place.

Soil types. The kind of soil that develops, and how quickly it develops, depends largely on the parent rock and the climate. Felsic rocks with high amounts of quartz weather to light-colored soil consisting of sand and clay with quartz grains. A basalt, on the other hand, which has a high percentage of ferromagnesian minerals, weathers to a darker-colored soil that is higher in clay content and lower in quartz.

Parent rocks form soils the most rapidly in wet, humid climates, where chemical weathering is enhanced. A **pedalfer** is a B horizon that is high in aluminum and iron. It develops in response to abundant rainfall, organic acids, and strong downward leaching. **Pedocals** are formed in arid climates by water drawn upward by subsurface evaporation and capillary action, which often crystallizes hard calcium salts in the soil. Pedocals are thin and poorly leached and have only a small amount of humus. **Alkali soils,** such as the white salt flats in the western United States, are pedocals that are toxic to plants because of the high salt content. **Hardpan** is a layer of soil, usually the B horizon, that is so hard (usually cemented by calcite or quartz) that it must sometimes be blasted loose. Common in the western United States, a hardpan formed by the precipitation of salt by evaporation is called **caliche.**

Laterites are highly leached, residual soils that form in tropical regions. The high temperature and rainfall result in deep and intense weathering. The laterite is typically bright red and composed of iron and aluminum oxides that are the most resistant to leaching. Valuable metals such as aluminum, copper, silver, gold, nickel, and iron are concentrated in laterites through **secondary enrichment** and can sometimes be mined economically. Secondary enrichment works in two ways: It can remove many of the other elements from the rock, which enriches the remaining valuable elements in place, or it can leach valuable elements such as copper and gold and deposit them at a particular level lower in the laterite.

Mass wasting is a natural result of weathering on slopes. Simply put, gravity pulls loose rock and soil downhill. **Mass wasting** is the process of erosion whereby rock, soil, and other earth materials move down a slope because of gravitational forces. It proceeds at variable rates of speed and is largely dependent on the water saturation levels and the steepness of the terrain. A destructive, rapid mass-wasting event is called a **landslide;** if the movement is slow enough that it cannot be seen in motion, it is called **creep.**

Three kinds of movement are generally recognized: flow, slip, and fall. A mass-wasting event is called a **flow** if the mass moves downslope like a viscous fluid. If the mass moves as a solid unit along a surface or plane, it is called a **slip.** A slip that moves along a surface parallel to the slope is called a **slide.** If the movement occurs along a curved surface where the downward movement of the upper part of the mass leaves a steep scarp (cliff) and the bottom part is pushed outward along a more horizontal plane, it is called a **slump.** Earth material that free-falls from a steep face or cliff is termed a **fall.**

Mass-Wasting Controls

A variety of conditions affect the development of mass wasting in a particular area. Steep slopes, widely varying altitude ranges (relief), the thickness of the loose earth material, planes of weakness parallel to the slopes, frequent freezing and thawing, high water content in the earth material, dry conditions with occasional heavy rainfall, and sparse vegetation are the factors that contribute to the unstable conditions that result in mass wasting. Movements can be triggered by the motion of earthquakes or too much weight added to the upper part of a slope, such as snowpack.

Angle of repose. The **angle of repose** is the steepest angle at which loose material will remain in place. It is largely dependent on the size, shape, and roughness of the particles. The angle varies from about 25 degrees to about 40 degrees. If the angle is exceeded by additional sedimentation or tilting, a slide or disturbance will result.

Gravity and friction. Rock particles and soil move downslope because of the forces of gravity. The gravity that acts on an object is a combination of the normal force and the shear force. The **normal force** is perpendicular to the slope the object rests on, and the **shear force** is parallel to the surface of the slope (Figure 26). Steep slopes have high shear forces; the steeper the slope, the greater the chance an object will slide. Friction, such as that from a rough bedrock surface, counteracts shear force. Rough, angular particles maintain steeper slopes than smooth rounded particles do. Water acts as a lubricant and reduces the force of friction, increasing the tendency to slide.

Gravitational Forces That Affect Mass Wasting

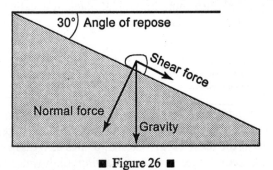

■ Figure 26 ■

The **shear strength** is an object's resistance to movement that needs to be overcome in order to make it move. Shear strength is proportional to how solid the mass is, the density of plant roots, the amount of water present, and the roughness of the particles in the mass.

The effects of water. In addition to acting as a lubricant, water increases the weight of a mass of earth material. Water reduces the shear strength by forcing sediment particles apart through **pore pressure,** which reduces the friction between the particles. Alternatively, smaller amounts of water that don't completely fill the pores are distributed as thin films around the sediment particles, which are attracted to each other through **surface tension,** increasing the cohesiveness of the mass. Thus, saturated materials are much more likely to flow than a mass that is only a little wet.

Types of Mass Wasting

Rockfalls and rockslides. Rockfalls occur when pieces of rock break loose from a steep rock face or cliff. These result from the rock face being undercut by rivers or wave action. Frost wedging may also eventually loosen large blocks, causing them to fall. The accumulation of rock debris at the base of a steep slope is called **talus.**

Rockslides usually follow a zone of weakness, such as a bedding plane or foliation plane. Separation of the rock is more likely along these planes because of their reduced shear strength. Water also tends to be channeled along these planes, which increases slippage. Collisions down the slope generally break the rock mass into rubble that eventually comes to rest. If steep slopes are involved, a fast-moving **rock avalanche** may result. The rockslide or rock avalanche loses energy and speed as it moves across more level terrain.

Debris flows. Debris flows are defined as mass-wasting events in which turbulence occurs throughout the mass. Varieties of these are called earthflows, mudflows, and debris avalanches.

When earth material moves down a hillside as a fluidlike mass, it is called an **earthflow.** These flows typically occur in humid areas on steep slopes with thick, clay-rich soil that becomes saturated with water during storms. The earthflow usually leaves a steep scarp behind where it separated from the hillside. Earthflows can be fast (a few hours) or slow (a few months). Velocities range from 1 millimeter per day to meters per day. Intermittent activity can continue for years as the earthflow continues to settle and stabilize. Earthflows typically have rounded, hilly fronts. A common trigger for an earthflow is the undercutting of the slope by erosion from wave action or rivers or by construction projects.

A variety of earthflow called **solifluction** is the flow of water-saturated earth material over an impermeable surface such as permafrost. It occurs frequently in bitterly cold regions such as in Alaska or Canada. Springtime temperatures thaw only the first few feet of the frozen ground (the active layer), which becomes saturated quickly and slowly flows over the ever-frozen permafrost below. Solifluction can occur on even the gentlest of slopes. Not forceful enough to break apart the surface vegetation, the migrating material drags it along like a wrinkled green rug. The soil finally settles on level ground at the base.

A **mudflow** is a liquidy mass of soil, rock debris, and water that moves quickly down a well-defined channel. Generally viscous and muddy colored, it can be powerful enough to move large automobiles and buildings. Mudflows occur most often in mountainous semiarid environments with sparse vegetation and are triggered by heavy rainfall that saturates the loose soil and sediment. They are also the natural result of volcanic ash build-ups on flanks of volcanoes and of forest fires that have exposed the soil to rapid erosion. A mudflow originating on a volcanic slope is called a **lahar.**

The deadliest variety of debris flow is the **debris avalanche,** a rapidly churning mass of rock debris, soil, water, and air that races down very steep slopes. It has been theorized that trapped air may increase the speed of an avalanche by acting as a cushion between the debris and the underlying surface.

Creep. A slow, gradual movement of soil or regolith downhill over time is called **creep.** Velocities are typically less than a centimeter per year. Freezing and thawing contribute to soil creep by progressively moving soil particles down the hill. Creep is manifested at the surface by such things as tilted utility poles that become more out of alignment every year. Vegetation helps reduce the rate of soil creep.

Slump. Earth material that has moved as a unit along a curved surface is called **slump**. A slumped mass of sediment is typically clay rich. Slump usually results when the geometrical stability of a slope is compromised by the undercutting of its base, such as by wave action, a meandering river, or construction.

Prevention of Mass Wasting

Proper design during construction projects can eliminate the potential for increased mass wasting. Human activities such as undercutting the base of the slope, adding weight to the upper part of the slope by building large structures, removing vegetation, and saturating the ground with water increase the risks of mass wasting. Engineering solutions include barriers and retaining walls, drainage pipes, terracing the slope to reduce the steepness of the cuts, and immediate revegetation. Rockfalls can be controlled or eliminated by the use of rock bolts, cables, and screens and by cutting back slopes to lesser gradients

The **hydrologic cycle** is the constant circulation of the earth's water through precipitation, evaporation, and transpiration (the release of water into the atmosphere by plants). It is the continuous exchange of water between the atmosphere, land, and ocean. **Running water** is the most active landscape-transforming agent on the earth's surface. Waterways erode, transport, and deposit rock and sediment to produce landforms such as canyons, valleys, deltas, alluvial fans, and floodplains.

Types of Water Flow

Streams (any flow of water within a natural channel regardless of size) are the most important kinds of **channel flow** that affect landscapes. A stream's **headwaters** are where the stream originates, usually in the higher elevations of mountainous terrain. The stream flows downhill and across lower elevations to its terminus, where it enters another stream, lake, or ocean. This terminus is called the **mouth** of the stream.

The stream is often flanked on both sides by a flat **floodplain** that is created when periodic flooding deposits mud and silt over extensive, low-lying areas. Flooding results when a stream's flow is increased and exceeds the capacity of the **stream channel.** Water sometimes moves overland during heavy storms as a **sheetwash,** a thin layer of unchanneled water. Sheetwashes typically occur in arid climates or where the ground is saturated and cannot accept any more water. Eventually sheetwash flow forms small channels called **rills;** rills join to form larger temporary streams.

About 80 percent of all rainfall is soaked into the ground and becomes groundwater or is taken up by plants and returned to the atmosphere through transpiration. Very heavy amounts of rainfall over short periods create **flash floods.** The flooding is a result of the ground becoming saturated and not being able to absorb any more water or of the water just coming too fast to be entirely soaked into the ground. Flash floods are common in the southwestern United States where the terrain is dry, rocky, and sparsely vegetated.

A **drainage basin** is the land area that contributes water to a stream and its tributaries (the smaller streams that flow into it). The size of the drainage basin depends on the size of the stream—large river systems have drainage basins that cover thousands of square miles. A small tributary to the Mississippi River, on the other hand, might have a drainage basin of only a few square miles.

The line of highest elevation that separates one drainage basin from another is called a **drainage divide.** The **Continental Divide** is a north-south line in the western United States and Canada that separates those streams that flow into the Pacific Ocean from those that empty into the Atlantic Ocean or the Gulf of Mexico.

Stream Dynamics

Stream gradient. The **stream gradient** is the downhill slope of the channel. For example, a gradient of 10 feet per mile means that the elevation of the channel drops a total of 10 feet over 1 mile of horizontal distance traveled. Gradients are typically the lowest at a river's mouth, and highest at its headwaters. The higher the gradient, the faster the stream flows.

Channel shape and texture. The shape and roughness of the channel also affect the velocity of the flowing water. A narrow channel that is V-shaped or semicircular in cross section results in faster flow; a wide, shallow channel yields a slower flow because there is more friction between the water and the stream bed. A smooth channel offers less friction than a rocky or boulder-strewn channel, resulting in faster flow. Thus a stream's velocity is greatest in a narrow, deep, smooth, and semicircular channel.

Stream velocity. The speed at which a stream flows is called the **stream velocity.** A fast river moves at a rate of about 5 miles per hour. The water moves most rapidly in the middle of the channel, where the water is deepest and friction is minimal. The water moves at a slower rate along the bed of the channel and the banks, where contact with rock and sediment (and therefore friction) is greatest. The greater the velocity of a stream, the greater its capacity to erode and transport earth materials over longer distances.

Stream discharge. A stream's **discharge** is the amount of water that flows past a certain point in a given amount of time. Discharge is usually expressed in cubic feet per second and represents the product of the cross-sectional area of the stream and the velocity:

$$\text{Discharge (cf/s)} = \frac{\text{channel}}{\text{width (ft)}} \times \frac{\text{average}}{\text{channel}} \times \frac{\text{average}}{\text{velocity (ft/s)}}$$

Discharge generally increases downstream because of additional water that is contributed from tributaries and groundwater that enter the main channel of flow. Stream discharges vary according to seasonal and precipitation changes. The rates of flow, discharge, erosion, sedimentation, transportation, and deposition increase dramatically during flooding and may be a hundred times greater than normal rates.

Stream Erosion

Streams are one of the most effective surface agents that erode rock and sediment. Erosional landscapes such as the Grand Canyon have been formed by constant erosion from running water over millions of years. In addition to eroding the bedrock and previously deposited sediments along its route, a stream constantly abrades and weathers the individual rock and soil particles carried by its water. Hydraulic action, abrasion, and solution are the three main ways that streams erode the earth's surface.

Hydraulic action. The ability of flowing water to dislodge and transport rock particles or sediment is called **hydraulic action.** In general, the greater the velocity of the water and the steeper the grade, the greater the hydraulic action capabilities of the stream. Hydraulic action is also enhanced by a rough and irregular stream bottom, which offers edges that can be "grabbed" by the current and that create uplifting eddies.

Abrasion. **Abrasion** is the process by which a stream's irregular bed is smoothed by the constant friction and scouring impact of rock fragments, gravel, and sediment carried in the water. The individual particles of sediment also collide as they are transported, breaking them down into smaller particles. Generally the more sediment that a stream carries, the greater the amount of erosion of the stream's bed. The heavier, coarser-grained sediment strikes the stream bed more frequently and with more force than the smaller particles, resulting in an increased rate of erosion.

Circular depressions eroded into the bedrock of a stream by abrasive sediments are called **potholes.** The scouring action is greatest during flood conditions. Potholes are found where the rock is softer or in locations where the flow is channeled more narrowly, such as between or around boulders.

Solution. Rocks susceptible to the chemical weathering process of **solution** can be dissolved by the slightly acidic water of a stream. Limestones and sedimentary rock cemented with calcite are vulnerable to solution. The dissolution of the calcite cement frees the sedimentary particles, which can then be picked up by the stream's flow through hydraulic action.

Sediment Load

The majority of a stream's sediment load is carried in solution (dissolved load) or in suspension. The remainder is called the bed load.

Dissolved load. Earth material that has been dissolved into ions and carried in solution is the **dissolved load.** It is usually contributed by groundwater. Common ions are calcium, bicarbonate, potassium, sulfate, and chloride. These ions may react to form new minerals if the proper chemical conditions are encountered during flow. Minerals may also precipitate in trapped pools through evaporation.

Suspended load. The **suspended load** is the fine-grained sediment that remains in the water during transportation. For example, a flooding river is muddy and discolored from the large amounts of sediment suspended in the water. The suspended load is generally made up of lighter-weight, finer-grained particles such as silt and clay. Most of the sediment in a stream is carried as suspended load. It does not contribute greatly to stream erosion, since it is not in frictional contact with the stream bed.

Bed load. The heavier, coarser-grained earth material that travels along the bottom of the stream is the **bed load. Traction** occurs when these fragments move along by rolling and sliding. Turbulent or eddying currents can temporarily lift these larger grains into the overlying flow of water—the grains advance by short jumps or skips until the surge diminishes and then fall back to the bottom because of their greater weight. This process is called **saltation.**

Capacity and competence. The maximum load of sediment that a stream can transport is called its **capacity.** Capacity is directly proportional to the discharge: the greater the amount of water flowing in the stream, the greater the amount of sediment it can carry. A stream's **competence** is a measure of the largest-sized particle it can transport; competence is directly proportional to a stream's velocity, which can vary seasonally. Because of increased capacity and competence, a single flood event can cause more erosion than a hundred years of standard flow.

Stream Deposition

A stream's sediment load is typically deposited, eroded, and redeposited many times in a stream channel, especially during climatic variations such as flooding. Sediments are deposited throughout the length of the stream as bars or floodplain deposits. At the mouth of the stream, the sediments are usually deposited in alluvial fans or deltas, which represent a lower-energy, more "permanent" depositional environment that is less susceptible to changes in the stream flow.

Bars. **Bars** form in the middle of the channel or along the banks of a stream at points where the velocity decreases, resulting in the deposition of some of the sediment load. Bars are ridges generally made up of gravel- or sand-sized particles. A subsequent flood event will erode bars, transport the sediments, and redeposit the material as a new bar farther downstream.

Floodplains. **Floodplains** are level strips of land on the sides of a channel that consist of fine-grained silt and clay deposited during episodes of flooding. Higher ridges of sand and silt called **natural levees** are deposited near the edge of the channel. As the water spreads outward from the channel, it loses energy and carries less sediment. The poorly drained and marshy areas behind the levees are called **backswamps.**

Alluvial fans. **Alluvial fans** are similar to deltas and are large fan-like accumulations of sediment that form where streams emerge from rugged terrain onto a broad, flatter surface. Stream velocities fall quickly, and the fan is built by continual braided stream activity. Large fans show graded patterns in which the coarsest-grained materials are deposited at the canyon mouth and the finer-grained materials spread outward in a fan shape.

Deltas. Sediment deposited at the mouth of a stream usually forms a thick, roughly wedge-shaped accumulation called a **delta**, the widest part of which is farthest from the stream mouth. **Distributaries** are dendritic, shifting channels that spread out across the delta from the main river channel and disperse the sediment load. Sediments on the delta's forward slopes are constantly shaped by water and wind action and redeposited by lake or ocean currents.

Topset beds are nearly horizontal layers of sediment deposited by the distributaries as they flow away from the mouth and toward the delta front. They overlie the sandy **foreset beds** that compose the main body of the delta, which dip downward from 5 to 25 degrees. These represent the gradual accumulations of sediment deposited over the forward slopes of the delta as it builds progressively outward into the receiving body of water. The **bottomset beds** are the finest silt and clay particles that are carried out into the deeper water or slide down the delta front into the deeper water.

Braided streams. A **braided stream** is one in which the water has lost its main channel and flows through a wandering network of rivulets around sandbars. Braided streams typically have wide channels. The braiding generally results from a flood-deposited midchannel bar that splits the flow. The water is diverted to the sides and erodes stream banks, widening the channel. Streams will usually be braided if they have high bed loads and easily erodible banks. The distributaries in a delta are also braided.

Meanders and oxbow lakes. The course of a stream bed can be continuously affected by erosion on the outside of a curve and deposition on the inside. This process will transform a gentle curve into a hairpinlike **meander.** Meanders continuously change location as they swing back and forth across a valley or migrate downstream over time. An **oxbow lake** is formed when a meander begins to close on itself and the stream breaks through and bypasses the meander (Figure 27). The cut-off meander is dammed by sedimentary deposits in the new channel—resulting in a body of water that is shaped roughly like a U (the shape of an oxbow, a piece of wood used to harness an ox). Oxbow lakes mark the location of former stream channels.

A Meandering Stream and Oxbow Lake

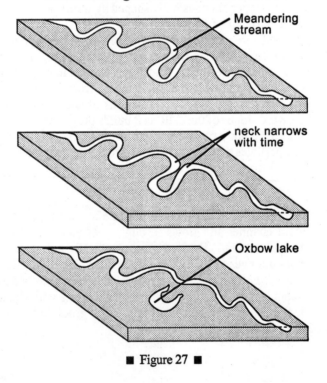

Meandering stream

neck narrows with time

Oxbow lake

■ Figure 27 ■

Stream Valleys

The erosion and transport of rock and sediment by a stream defines the shape and extent of its **valley.** V-shaped valleys and wide valleys with flat floors are the most common varieties.

Downcutting. A valley is the result of **downcutting,** whereby a stream's channel erodes directly downward. As downcutting continues, erosion and mass wasting begin to work on the exposed, vertical sides of the channel, eroding them into slopes and widening the valley (Figure 28).

A Downcutting Stream Profile

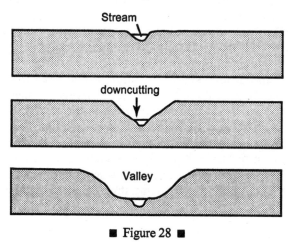

■ Figure 28 ■

Slot canyons, vertical-walled rock canyons where mass-wasting processes have been very limited, are a common feature in the western United States (Figure 29).

Slot Canyons

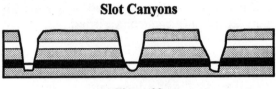

■ Figure 29 ■

As a stream flows downslope and gains more water from tributaries, the valley becomes wider because of greater mass wasting. Downcutting proceeds until the **base level** is reached—the elevation of the most horizontal flow and lowest velocity. For streams that empty into the ocean, base level is essentially sea level. Base level for continental streams is generally the lowest elevation of the valley.

Ungraded and graded streams. An **ungraded stream** is one that is still actively downcutting and smoothing out its irregular gradient through erosion. It is characterized by rapids and waterfalls. In contrast, a **graded stream** has smoothed out its longitudinal profile to resemble a smooth, concave-upward curve (Figure 30). How long it takes a stream to become graded is influenced by sediment load, which affects the rates of erosion and downcutting. Dams also have a major impact on grading by reducing stream flow and sediment load.

An Ungraded and a Graded Stream

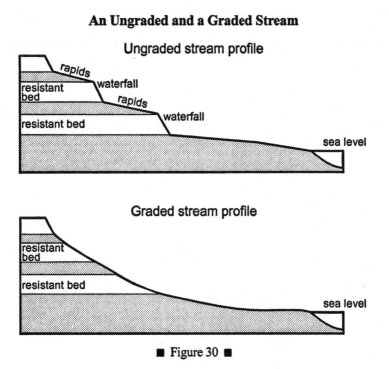

■ Figure 30 ■

Headward and lateral erosion. Valleys are further developed by headward erosion and lateral erosion. **Headward erosion** results when a valley is extended upward above its original source by gullying, mass wasting, and sheetwash flow. **Lateral erosion** occurs when the stream meanders or braids back and forth across its valley floor or channel, undercutting and eroding its banks. This results in mass wasting of the gradually more unstable slopes and forms a wider floodplain.

Stream terraces. **Stream terraces** are steplike benches that occur above the stream bed and floodplain. They are cut into bedrock or are remnants of older river sediments that have since been eroded. Terraces form in response to flooding or changes in base level.

Incised meanders. **Incised meanders** are steep-walled canyons that result from the downcutting of a meandering stream. Usually without floodplains, they are thought to be the result of the uplift of a meandering stream above its base level.

Regional Erosion

Slope erosion. Early theories suggested that mountains gradually eroded down to form plains over very long stretches of time—the sides and tops of the mountains were eroded to gentler slopes at similar rates, gradually losing elevation and approaching base level. Today it is more accepted that many slopes erode according to **parallel retreat.** Instead of eroding to gentler angles over time, slopes maintain their original steepness as they erode progressively back from river channels (Figure 31).

Parallel Retreat

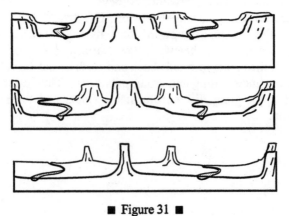

■ Figure 31 ■

The climate, rock type, and rock structures are the most important factors in slope erosion. As previously noted, rocks of different compositions weather at different rates—harder rocks form steeper slopes and generate coarser-grained, more angular talus slopes that are resistant to weathering and erosion. Sand and clay slopes erode more easily and are less steep. Certain climates accelerate soil creep and chemical weathering, which result in more rounded topography. Rock structures such as folds, bedding planes, and faults can also affect the slopes that result from weathering.

Drainage patterns. A stream and its tributaries form geometrical arrangements called **drainage patterns.** A drainage pattern can be greatly influenced by the geologic formations through which it passes. A treelike **dendritic pattern** develops in a rock type that erodes uniformly, such as granite. A **radial pattern,** which resembles the spokes on a wheel, occurs when the streams originate on the flanks of conical mountains. A **trellis pattern** consists of parallel main streams with short tributaries on either side that form in areas of tilted sedimentary rocks that create parallel ridges and valleys. Bedrock that is regularly fractured or jointed in 90-degree angles can create **rectangular patterns,** which have distinctive right-angle bends.

At times in the geologic past, up to a third of the earth's surface was covered by thick sheets of glacial ice. The last glacial ice sheet in North America had melted by about 10,000 years ago. The movement of glaciers across the continents profoundly changed the landscape through extensive erosion, transportation, and deposition of rock and sediment. Glaciers sculpted some of the spectacular erosion features in the high mountain ranges and also eroded large areas down to flat bedrock surfaces. The deposition of glacial sediments formed some of the extensive, gently rolling landscapes we see today, such as in southern Wisconsin.

Glaciers are thick, sprawling masses of ice that form on land during cooler climatic periods. Glaciers begin to grow when more snow accumulates than is lost through melting during the year. The older snow compacts, recrystallizes, and turns to ice from the increasing weight of the new snow above it. When a glacier reaches a sufficient size and mass, the force of gravity begins to move it downslope.

Glaciation is the movement of an ice sheet over a land surface. Two types of glaciation are recognized: continental and alpine. **Continental glaciation** affects a broad section of a continental land mass, such as Antarctica. **Alpine glaciation** is usually restricted to deep valleys in high mountainous terrain.

The idea that, compared to what we see today, a much larger portion of the earth was covered by glaciers in the geologic past was first proposed in Europe in the early 1800s by Swiss scientist Louis Agassiz. Known as the **theory of glacial ages,** it also meant that colder climates had existed for thousands of years. A series of glacial episodes, separated by periods of warmer climate, have occurred in the last few million years. Glaciers during this "Ice Age" stripped the soil from the northern part of North America (Canada) and deposited it farther south, creating the rich farmland in the midwestern United States.

Types of Glaciers

Glaciers can be found in both polar and more temperate climates. They are the most abundant in the polar regions, where it remains so cold that only a minor amount of water is lost through melting or evaporation. They can also be found in the highest mountains in temperate or even tropical latitudes where temperatures remain cold throughout the year, such as in the Pacific Northwest of the United States and Canada, Alaska, and South America. More snow and ice accumulate during the winter months in these mountain ranges than is lost as meltwater in the summer.

About one-tenth of the land surface on Earth is covered by glaciers today. Over 75 percent of this amount is on Antarctica, and 10 percent is on Greenland. The remainder occurs in mountain regions across the world. If the entire Antarctic ice sheet melted, it would raise the sea level about 60 meters (200 feet) and flood many cities in low-lying coastal areas around the world.

Ice sheets are associated with continental glaciation and cover large areas of a landmass (over tens of thousands of square kilometers). Ice sheets exist in Greenland and Antarctica. **Ice caps** are similar to ice sheets but are much smaller—they are usually found in the highest part of a mountain range, where the snow accumulation is the greatest. An ice cap can be a source for multiple valley glaciers.

Valley glaciers (or **alpine glaciers**) are masses of ice that are restricted to high mountain valleys. As they move downslope, they can connect with larger valley glaciers. The majority of alpine glaciation is the result of the repeated advance and retreat of valley glaciers. Valley glaciers are common in the mountain ranges of the United States and Canada. **Piedmont glaciers** are the forwardmost extension of valley glaciers and form where the ice emerges at the front of the mountain range. The ice spreads out on the flat terrain to form a wide sheet at the mouth of the valley.

How Glaciers Develop

Snow into ice. Snow turns into glacial ice because of the pressure from overlying layers of snow. The increased weight compacts the delicate snowflakes and collapses pockets of air. The snowflakes become rounded granules called **firn,** which are held loosely together by new ice that acts as a cement. The greater the overlying weight, the greater the amount of compaction and recrystallization that leads to the development of thick slabs of glacial ice.

Wasting and calving. As its mass increases, a glacier begins to move downslope, or **flow,** under the influence of gravity. It commonly picks up loose rock and sediment or breaks off pieces of the irregular rocky surface. When glacial ice reaches its point of farthest advance, it is **wasted,** or **ablated,** through melting. A small amount of the ice evaporates directly into the atmosphere from the warmed surface of the glacier. Large blocks of ice may break off, or **calve,** from the glacial face and plunge into the water of a lake or ocean as an **iceberg.** In extremely cold climates, most glaciers lose their ice through calving.

Budget. A glacier's **budget** is defined as the ratio between ice gained and ice lost. When a glacier gains more volume from new snowfall than it loses from melting, it has a **positive budget.** This positive growth is reflected by the outward or downslope movement of an **advancing glacier** because of the increased snow mass at the top, even if the front of the glacier is melting. A glacier with a **negative budget** loses more volume than it gains and is therefore a **receding glacier.** A receding glacier may at times still move downslope but cannot in total overtake its more rapid rate of uphill recession from melting. A glacier that has a **balanced budget** neither advances nor recedes.

Zone of accumulation and zone of wastage. The upper elevations of a glacier that are perennially covered in snow are called the **zone of accumulation.** The lower portion of the glacier where the ice is lost is called the **zone of wastage.** The **snow line** is the irregular boundary between these two zones. The position of the snow line varies according to climatic variations and the glacier's budget. A snow line that moves down the glacier indicates the glacier has a positive budget; a snow line that moves up the glacier reveals a negative budget.

The terminus. The front of the glacier, or its **terminus,** moves down the valley if a glacier has a positive budget; the reverse is true if the glacier has a negative budget. Colder temperatures are not the only reasons a glacier extends forward. Other causes could be that conditions are wetter and more snowfall is accumulating during the winter months or that the summer is cloudier and cooler. Alternatively, a retreating glacier could mean that the winter months have been drier but just as cold, resulting in less snowpack, or that a sunnier summer has resulted in more wastage. Experts guess that a worldwide decrease in the mean annual temperature of only 4 or 5 degrees centigrade could trigger the onset of another glacial period.

Glacier Movement

Movement of ice sheets. An ice sheet moves downslope in a number of directions from a central area of high altitude and is not restricted to a channel or valley. The ice sheet must expand because of the constant accumulation of ice and snow. Ice sheets do not move as quickly as alpine glaciers because there is less slope and more mass involved. Ice sheets move mostly by plastic flow. Mountain ranges are completely buried by the ice sheet at the South Pole, which is greater than 3,000 meters thick.

Movement of valley glaciers. Glaciers can move more than 15 meters a day. The larger volumes of ice on steeper slopes move more quickly than the ice on the more gentle slopes farther down the valley. These dynamics allow a glacier to replenish the ice that is lost in the zone of wastage. Glaciers in temperate zones tend to move the most quickly because the ice along the base of the glacier can melt and lubricate the surface. Other factors that affect the velocity of a glacier include the roughness of the rock surface (friction), the amount of meltwater, and the weight of the glacier.

Basal sliding and plastic flow. A valley glacier has various components of flow. First, the entire glacier moves as a single mass over the underlying rock surface. The pressure from the weight of the glacier generates a layer of water that helps the ice glacier move downslope. This process is called **basal sliding.**

In addition to basal sliding, which slowly moves the glacier downslope as a unit, **plastic flow** causes glacial ice buried underneath more than about 50 meters to move like a slow-moving, plastic stream. The central and upper portions of a glacier, as do those portions of a stream, flow more quickly than those near the bottom and sides, where friction between the ice and valley walls slows down the flow. In general, the rate of plastic flow is greater than the rate of basal sliding.

Above a depth of about 50 meters, the weight of the overlying ice is not sufficient to cause plastic flow. This more rigid upper zone, which is called the **zone of fracture,** is carried along the top of the plastic flow piggyback style. Sometimes the zone of fracture moves faster than the underlying plastic flow. When this happens, especially down a steep slope, the surface breaks into a series of deep fissures called **crevasses.** Crevasses also result where a valley glacier curves because the ice flows faster toward the outside of the curve than the inside. A steep, rapid descent may result in an **icefall,** a piled-up mass of splintered ice blocks from a series of rapidly formed crevasses.

Glacial Erosion

Basal sliding erodes the rock surface underlying a glacier. Meltwater freezes in cracks, and pieces of bedrock are pried loose and incorporated into the ice (similar to frost wedging). These pieces of rock grind against each other and the bedrock underneath. The intensity of the grinding is proportional to the weight and pressure of the glacier. **Faceted** rocks are those rocks along the base that are worn flat from erosion. The surfaces of the bedrock and the rock fragments are **polished** just like rocks in a rock tumbler are. Rocks embedded in the ice scratch long, deep grooves in the bedrock called **striations,** which indicate the direction of ice movement. The constant grinding of rocks in the glacier creates a fine-grained **rock powder** that turns the meltwater white when released by ablation.

Other erosional features associated with glaciers include increased mass wasting from rapid downcutting of valley sides, frost wedging that creates rockfalls, and avalanches that contribute additional rock and soil material.

Glacial Landforms

The most striking glacial erosional features are associated with alpine glaciation. In fact, rugged mountainous areas can be made even more spectacular by glacial action. Alpine glaciers transform V-shaped valleys made by streams into deeper U-shaped valleys called **glacial troughs**—the ice is too massive to follow the stream bed and pushes right through, scouring out a U shape. The ice also erodes away the ends of any ridges along the valley walls. These eroded ridges are called **truncated spurs.** The valleys of tributaries can also be truncated, forming **hanging valleys** that are higher than the main valley and often marked by waterfalls.

The mass of ice at the top of a glacial valley ultimately forms a steep-sided, circular hollow called a **cirque.** Mass wasting and frost wedging also contribute to the formation of a cirque. A **bergschrund** is a crevasse that forms where the glacier separates from the cirque wall and is commonly filled with rock fall debris. A **horn** is a sharply defined peak that has formed from erosional processes along the rim of the cirque. A steep ridge called an **arete** commonly extends downward from a horn to separate two adjacent glacial valleys.

An advancing glacier scours out a series of depressions in the underlying bedrock, which later fill with water and become **rock-basin lakes,** or **tarns.** Tarns are best developed in softer or highly fractured bedrock. Tarns are less common on smooth, hard bedrock surfaces because it is more difficult for the glacier to "grab hold" and break off pieces of rock.

Ice sheets, with their slower rates of movement and greater weight, tend to grind down and smooth out the irregular, or sometimes mountainous, underlying surface. Exposed bedrock is polished and striated. The rounded geologic landforms and extensive, flat, bedrock surfaces in Ontario, Canada, are good examples of how an ice sheet affects the surface.

Glacial Deposits

Load. An advancing ice sheet carries an abundance of rock that was plucked from the underlying bedrock; only a small amount is carried on the surface from mass wasting. The rock/sediment load of alpine glaciers, on the other hand, comes mostly from rocks that have fallen onto the glacier from the valley walls. The various unsorted rock debris and sediment that is carried or later deposited by a glacier is called **till.** Till particles typically range from clay-sized to boulder-sized but can sometimes weigh up to thousands of tons. Boulders that have been carried a considerable distance and then deposited by a glacier are called **erratics.** Erratics can be a key to determining the direction of movement if the original source of the boulder can be located.

Features left by valley glaciers and ice sheets. Moraines left by valley glaciers are shown in Figure 32, and features left by a receding ice sheet are shown in Figure 33. **Moraines** are deposits of till that are left behind when a glacier recedes or that are carried on top of alpine glaciers. **Lateral moraines** consist of rock debris and sediment that have worked loose from the walls beside a valley glacier and have built up in ridges along the sides of the glacier. **Medial moraines** are long ridges of till that result when lateral moraines join as two tributary glaciers merge to form a single glacier. As more tributary glaciers join the main body of ice, a series of roughly parallel medial moraines develop on the surface of main glacier.

Moraines

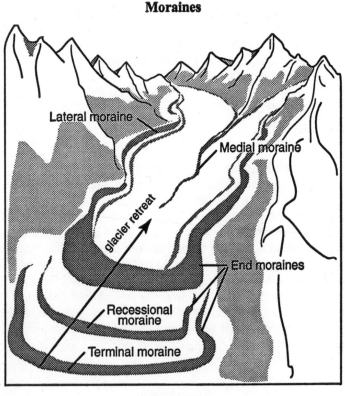

Lateral moraine

Medial moraine

glacier retreat

End moraines

Recessional moraine

Terminal moraine

■ Figure 32 ■

An extensive pile of till called an **end moraine** can build up at the front of the glacier and is typically crescent shaped. Two kinds of end moraines are recognized: terminal and recessional moraines. A **terminal moraine** is the ridge of till that marks the farthest advance of the glacier before it started to recede. A **recessional moraine** is one that develops at the front of the receding glacier; a series of recessional moraines mark the path of a retreating glacier.

A thin, widespread layer of till deposited across the surface as an ice sheet melts is called a **ground moraine.** Ground moraine material can sometimes be reshaped by subsequent glaciers into streamlined hills called **drumlins,** long, narrow, rounded ridges of till whose long axes parallel the direction the glacier traveled.

As a glacier melts, till is released from the ice into the flowing water. The sediments deposited by glacial meltwater are called **outwash.** Since they have been transported by running water, the outwash

Landforms Produced After Ice Recession

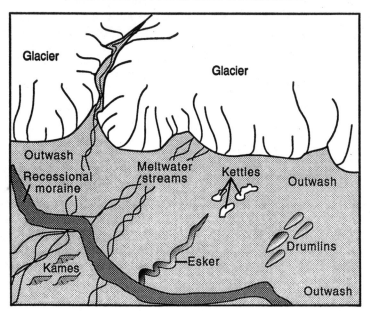

■ Figure 33 ■

deposits are braided, sorted, and layered. The broad front of outwash associated with an ice sheet is called an **outwash plain;** if it is from an alpine glacier it is called a **valley train. Kames** are steep-sided mounds of stratified till that were deposited by meltwater in depressions or openings in the ice or as short-lived deltas or fans at the mouths of meltwater streams. The rapid build-up of sediments can bury isolated blocks of ice. When the ice melts, the resultant depression is called a **kettle. Kettle lakes,** common in the upper Midwest of the United States, are bodies of water that occupy kettles.

Eskers are long, winding ridges of outwash that were deposited in streams flowing through ice caves and tunnels at the base of the glacier. Generally well sorted and cross-bedded, esker sands and gravels eventually choke off the waterway.

The great volume of meltwater often results in the formation of glacial lakes between the end moraines and the retreating glacier front. The sediments that form at the bottom of the lake consist of fine-grained silt and clay that have an alternating light-dark layering. A **varve** consists of one light-colored bed and one dark-colored bed that represent a single year's deposition. The light-colored layer is mostly silt that was deposited rapidly during the summer months; the dark layer consists of clay and organic material that formed during the winter. The age of a glacial lake can be determined from the number of varves that have formed on the lake bottom.

Glaciers in the Past

The earliest known glaciation occurred about 2.3 billion years ago and is recognized in Ontario, Canada, from older tills that have lithified into a rock called **tillite.** Tillites contain angular, unsorted rock fragments, many of which are polished, striated, or faceted. Another major period of glaciation occurred about 600 million years ago. Tillites from about 300 million years ago indicate that the ice sheet flowed over a supercontinent that later split apart to form Africa, Australia, Antarctica, India, and South America.

Glaciation has occurred more frequently in the last 20 million years. Episodic continental glaciations span back to about 3 million years ago. A number of different glacial periods occurred during the Pleistocene epoch, more commonly known as the "Ice Age." The last period of glaciation that covered a large portion of North America and Europe peaked about 18,000 years ago. Evidence for earlier glaciation is less complete because of weathering in the interglacial warm periods and subsequent glaciations. The estimated worldwide temperature difference between the Pleistocene and today is only about 5 degrees centigrade.

Geologists do not understand why glaciers form, advance, and retreat. Possible causes include variations in the earth's orbit and inclination to the sun, atmospheric changes, volcanic eruptions, changes in continental positions, changes in ocean currents, or movements in the Antarctic ice sheet. Serbian mathematician Milutin Milankovitch has shown that climatic variations in the past 100,000 years coincide with periodic variations in the amount of solar energy received by the earth.

North American Glaciation

The last major glaciations in North America during the Pleistocene covered all of Canada and the northern third of the United States. The thickest, central portion of the ice sheet covered Hudson Bay. The ice sheet stripped Canada of its topsoil, scoured and polished bedrock, and gouged out numerous future lake basins. The till and outwash were deposited to the south, forming the fertile farmlands of the United States. The ice carved out the Great Lakes basins, which are rimmed by end moraines. Glacial lakes were abundant in the Canadian prairies, North Dakota, and Minnesota. Alpine glaciers remained in California, the Rocky Mountains, and the northern Appalachians after the main ice sheet retreated.

The formation of the vast ice sheet lowered the sea level by about 130 meters (430 feet). Evidence for this includes terrestrial fossils, such as mammoths, that have been found beneath the ocean on the continental shelves. As glaciers receded, more water was contributed to the oceans and sea levels rose. A **fiord** is a steep-walled, fingerlike coastal inlet that was carved by glacial action and later flooded by the rising sea.

Pluvial lakes formed during the wetter climates that existed during and after glacial retreat. For example, the Great Salt Lake in Utah is a remnant of a much larger pluvial lake. Most pluvial lakes have shrunk because of the more arid conditions that have prevailed since they formed.

The tremendous weight of glacial ice may depress the earth's crust more than 200 meters. After the glacier retreats, the crust begins to move back to its previous level. This **crustal rebound** is still in progress around the Great Lakes in the United States.

Groundwater is extremely important to our way of life. Most drinking water supplies and often irrigation water for agricultural needs are drawn from underground sources. More than 90 percent of the liquid fresh water available on or near the earth's surface is groundwater. Hot groundwater can also be a source of energy. **Groundwater** is derived from rain and melting snow that percolate downward from the surface; it collects in the open pore spaces between soil particles or in cracks and fissures in bedrock. The process of percolation is called **infiltration.**

Porosity

The percentage of a rock or sedimentary deposit that consists of voids and open space is its **porosity**—the greater its porosity, the greater its ability to hold water. Sediments are usually more porous than rocks. Sedimentary rocks tend to be more porous than igneous rocks because there is more open space between the individual sediment grains than between the minerals in a crystallized rock. The porosity of loose sand is about 40 percent; compacting and dewatering the sand reduces the porosity to about 15 percent; the lithification of the sand into a sandstone rock by formation of cement between the sand grains reduces the porosity to about 5 percent or less.

Open space in fractures is also considered part of a rock's porosity. An igneous rock that is jointed, fractured, or shattered can hold as much water in its cracks as sedimentary rocks can hold between their grains.

Permeability

The ease with which fluid is transmitted through a rock's pore space is called **permeability**. Although a rock may be very porous, it is not necessarily very permeable. Permeability is a measure of how interconnected the individual pore spaces are in a rock or sediment. A sandstone is typically porous and permeable. Shales are porous but have a lower permeability because the finer grain size creates smaller pore spaces. Igneous rocks tend to have low porosity and low permeability unless they are highly fractured by tectonic processes.

The Water Table

Water flows downward through soil and bedrock because of the force of gravity. It continues in that direction until a depth of about 5 kilometers (3 miles) is reached, where porosity and permeability cease. The pore space above this level begins to fill progressively upward with groundwater.

The saturated zone. The rock and soil in which all the open spaces are filled with water is called the **saturated** (or **saturation**) **zone.** As the top of the saturated zone rises toward the surface, it reaches a level of equilibrium with the overlying unsaturated zone.

The unsaturated zone. The **unsaturated zone** (or **zone of aeration**) is the rock and sediment in which pore spaces contain mostly air and some water and therefore are not saturated. The unsaturated zone typically starts at the surface and extends downward to the saturated zone. The contact between the saturated and unsaturated zones is called the **water table** (Figure 34).

The Water Table

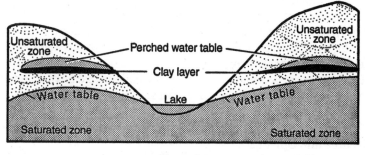

■ Figure 34 ■

There is "room" for air in the unsaturated zone because the water is held to the sides of the soil particles through the force of **surface tension**. Surface tension also causes water to rise up into the unfilled pore spaces from below through a process called **capillary action.** The lower part of the unsaturated zone that draws water upward from the water table is called the **capillary fringe,** which is usually only a few feet thick.

Perched water tables. A **perched water table** (Figure 34) is an accumulation of groundwater that is above the water table in the unsaturated zone. The groundwater is usually trapped above an impermeable soil layer, such as clay, and actually forms a lens of saturated material in the unsaturated zone. A perched water table is generally insufficient to supply domestic groundwater needs, and often runs dry after being drilled. If the perched water table intersects a sloping surface, it may be manifested by springs or seeps along the line of intersection.

Migration of groundwater. The movement of groundwater is generally slow and ranges from 1 inch to a 1,000 feet per day. In addition to moving vertically downward for hundreds of feet, it also flows laterally, roughly parallel to the slope of the surface of the water table.

The slope of the water table is generally proportional to the slope of the overlying land surface: the steeper the topography, the steeper the slope of the water table. The steeper the slope of the water table, the faster the groundwater flows. The groundwater also moves more quickly in those sedimentary or rock formations that have a higher permeability relative to other formations.

Streams and Springs

The dynamics of groundwater movement have an important effect on stream flow. Groundwater that migrates into the stream channel increases stream flow; water in a stream can also enter the unsaturated zone, reducing stream flow.

Gaining streams. A **gaining stream** (Figure 35) is one into which groundwater flows from the saturated zone. The channels of gaining streams are usually at or below the level of the water table. Bodies of water and marshes form when the water table intersects the land surface over a broad, fairly flat area.

A Gaining Stream

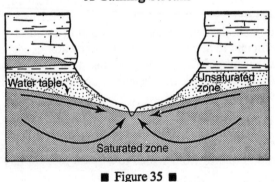

■ Figure 35 ■

Losing streams. The channel of a **losing stream** (Figure 36) lies above the water table and loses water into the unsaturated zone through which it is flowing This water then migrates down toward the water table. A losing stream can induce the local water table to rise. In drier climates a losing stream may actually disappear underground as its water content becomes progressively diminished downstream.

A Losing Stream

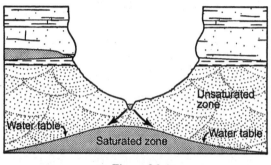

■ Figure 36 ■

Springs. A **spring** is a natural flow of groundwater from a rock opening that results when the water table intersects a sloping land surface. Springs can be seasonal—for example, during the wet season the saturated zone is closer to the surface because of increased rainfall, often resulting in more springs.

Aquifers. **Aquifers** are porous, permeable, saturated formations of rock or soil that transmit groundwater easily. The best aquifers are coarse-grained sediments such as sand and gravel. A **confined aquifer** is overlain by a less permeable bed that keeps the water in the aquifer under pressure; an **unconfined aquifer** does not have a confining bed that separates the zone of saturation from the unsaturated units above it. Impermeable formations such as shale, clay, or unfractured igneous rocks that retard water flow are called **aquitards** or **aquicludes.**

If more water is removed from an aquifer through pumping than is introduced through recharge, the water table drops. This often results in wells that go dry or a surface that sinks because the ground surface is no longer as supported. This subsidence inflicts expensive damage on buildings, roads, and pipelines. Heavy use of an aquifer can be balanced through **artificial recharge,** a process by which treated industrial wastewater or floodwaters are stored in infiltration ponds. The water soaks into the ground to replenish the groundwater or is pumped back into the aquifer.

Wells. Wells are drilled into the water table to tap aquifers for domestic, industrial, and agricultural use. The level of the water table fluctuates with changing climatic conditions. During a dry period, the water table drops to a deeper level because water has drained out of the saturated zone into the rivers. During wet periods the water table rises because of the additional water percolating down from the surface into the zone of saturation.

The water table tends to be closer to the surface in valleys than on hillsides. **Recharge** occurs in those areas where new water is added to the saturated zone and replenishes water that has been lost.

The best wells are drilled deeply enough to supply a continuous flow of water during all the seasonal changes in the water table—thus they draw water from below the lowest level of the water table. In **artesian wells,** which tap water from confined aquifers, either the water level in the well simply rises above the aquifer (**nonflowing artesian wells**) or it spouts at the surface (**flowing artesian wells**). Whether a well is flowing or nonflowing depends on the amount of pressure that is exerted on the groundwater in the confined aquifer.

When water is pumped from a well, the water table is generally lowered around the well. This local lowering of the water table is called **drawdown.** Centered on the stem of the well, it has the shape of an inverted cone called the **cone of depression.** The drawdown decreases with increasing distance from the well.

Effects of Groundwater Flow

The dissolution of calcite from limestone by slightly acidic groundwater results in the gradual widening of cracks and joints that may ultimately develop into a series of openings, or **caves.** Most caves develop below the water table. After the caves are elevated above the water table or when the water table drops, the water drains out and the caves become filled with air.

The groundwater that percolates through the cracks in the cave contains calcium and bicarbonate from the dissolution of limestone. As the water drips from the cave's ceiling, CO_2 gas is released and a small amount of calcite crystallizes where the drop is attached to the ceiling. More CO_2 is lost from the water when the drop hits the floor, causing more calcite to precipitate. By this process, stalactites and stalagmites form. **Stalactites** look like icicles that hang from cave ceilings; **stalagmites** are cone-shaped masses that build up on cave floors underneath dripping stalactites. A **column** results when stalactites and stalagmites grow long enough to join into one structure. A more general term for a deposit of calcite precipitated by dripping water is **dripstone,** and as a group, the varieties of dripstone found in caverns are called **speleothems.** Ribbony, sheetlike calcite deposits that are deposited by a thin film of water running over cave surfaces are called **flowstone.**

Caves become less structurally stable as greater amounts of limestone are eroded away. When a portion of a cave system collapses, it may create a **sinkhole,** or basinlike depression, at the surface. Sinkholes, which can develop suddenly and be large enough to "swallow" buildings and homes, are prevalent in states such as Florida, Missouri, Indiana, and Kentucky, which are underlain by abundant limestone. **Karst topography** is an irregular land surface dotted with numerous sinkholes and depressions related to underlying cave systems.

Groundwater that has a high concentration of silica is the primary agent in forming **petrified wood.** The groundwater soaks through the buried wood and precipitates silica in the porous organic structure, preserving the finest details. When silica or calcite layers precipitate from groundwater in a spherical cavity (usually in limestone), the often intricately layered mass that results is called a **geode**. A **concretion** is a mass of silica or calcite that precipitates around an organic nucleus, such as a leaf or fossil, in sedimentary rocks. Usually geodes and concretions are more resistant to weathering than the enclosing rock and stand out in weathered rock faces.

Groundwater Pollution

Sources of groundwater pollution. Because it is mixed and circulated over a large area, groundwater is relatively clean, but the increased population and industrialization of the twentieth century has led to serious groundwater contamination problems in many parts of the country. Farming contaminants include pesticides, herbicides, animal waste, and manure. A variety of contaminants from city and county dumps such as heavy metals (mercury, lead, chromium, copper, cadmium, arsenic) and other industrial compounds enter the groundwater from rainwater that has percolated through the landfill. Wastes from septic tanks, sewage plants, and slaughterhouses may also contribute dangerous bacteria and parasites to the groundwater. Industries frequently use radioactive compounds, cyanide, polychlorinated biphenyls, and a degreaser called trichloroethylene that are being found in increasingly greater amounts in groundwater. Gasoline and other fuel derivatives such as xylene and benzene are carcinogens that frequently enter the groundwater from leaking storage tanks. Old mining sites contribute mercury, cyanide, and heavy metals to the groundwater; smoke from old smelters contaminated soils for hundreds of square kilometers with metals such as lead, arsenic, and cadmium, which also migrated into the groundwater.

Contamination identification and cleanup. Most compounds form a **contamination plume** in the groundwater that grows wider as it spreads outward from the point of contamination, called the **point source.** If the plume is flowing through sand, a portion of the contaminants are naturally filtered from the groundwater. Even though the plume widens downgradient, the concentration of the contaminants tends to decrease through filtering, dilution, or the natural breakdown of substances over time and distance called **natural attenuation.**

A contamination plume is identified by drilling monitoring wells and routinely sampling the water for contaminants. A series of monitoring wells studied over time reveals details about the direction of groundwater flow and the level of contamination. Once the point source is identified, cleanup work includes removing contaminated material and soil at the surface and treating the groundwater. Groundwater is typically pumped out of the ground through a system of wells, cleansed, and pumped back into the aquifer. This procedure can last for thirty years or longer.

Geothermal Energy

Groundwater can be heated by a body of cooling magma or by penetrating deeply into the earth's crust along faults and being heated by the increased geothermal gradient.

Hot springs and geysers. Heated groundwater rises to the surface as hot springs and geysers. A **hot spring** consists of water 6 to 9 degrees centigrade warmer than the mean annual air temperature for the locality where it occurs. Hot-spring pools are often steaming and actively forming new minerals. The hot springs in Yellowstone National Park and Nevada are good examples of groundwater heated by igneous activity.

A **geyser** is a more explosive hot spring that periodically erupts scalding water and steam. **Fumaroles** are vents from which steam and other gases escape. Geyser eruptions result from newly formed mineral deposits that clog the throat of the vent or from accumulations of vapor bubbles that increase the internal pressure. Water temperatures are generally near boiling. The hot water rapidly cools at the surface and precipitates new minerals around the geyser vents.

Typically composed of calcite or silica, the build-up of these ledge-like layers is called a **sinter** around a hot spring and **geyserite** around a geyser. A **mudpot** is a vent that produces thick, boiling mud and sulfurous gases. Rich precious-metal deposits such as gold and silver are often associated with hot spring activity.

The use of geothermal energy. **Geothermal energy** is the energy produced when heated groundwater is tapped by wells and used to generate electricity. Although geothermal energy is one of the cleanest forms of energy, hydrogen sulfide gases and other toxic compounds may be associated with it. Compared to that of other sources of energy, the use of geothermal energy is not widespread; it can, however, be locally important, as it is in Iceland.

The dominant agents of erosion in coastal environments are waves. Driven by wind and tidal action, waves continuously erode, transport, and deposit sediments along ocean coastlines. The sand is also continuously moved parallel to the beach by longshore currents and is frequently deposited in harbors, where it must be periodically dredged to keep the harbor open for commercial shipping.

Waves move because the surface of the water gains energy from winds that blow over it. Short waves tend to be produced by local storms; long rolling swells are generated by large, distant storm systems up to thousands of miles away. The **surf** is that zone where waves break against the shoreline. The energy of the waves then sorts the sand and moves it along the beach. Beaches expand or dwindle according to changing coastal conditions. The restless ocean and its waves make many coastal landforms fragile and short lived. Wave energy is dependent on weather conditions, the length of the waves, wind speed, the duration of the wind, and the distance the wind travels over open water (**fetch**).

Waves

Wave height. The tops of waves are called **crests** and are separated by the lowest points, called **troughs.** The most powerful waves have the greatest wave height. Wave height is the vertical distance between the crest and the trough. Normal waves can be nearly 5 meters (15 feet) high; severe tropical storms can generate waves up to 15 meters (50 feet) high.

In rare cases, wave energy is derived from a submarine earthquake. Called a **tidal wave, seismic sea wave,** or **tsunami,** these gigantic walls of water can be as high as 90 meters (300 feet) and do tremendous damage to coastlines and cities.

Wavelength. The **wavelength** is the horizontal distance between two adjacent crests or two adjacent troughs (Figure 37). Typical wavelengths vary from about 30 to 300 meters (100–1,000 feet), and waves move at speeds up to 50 miles per hour. The depth of the wave motion is about half the wavelength; for example, if the wavelength is about 150 meters, 75 meters below the wave crest the water is calm.

Waves

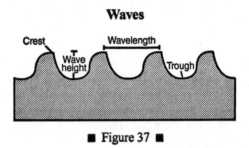

■ Figure 37 ■

Wave movement. Individual molecules of water are not physically transported with the waves as they move across the surface. The energy of the wave passes *through* the water molecules and does not carry them along. At the surface of the wave, a water particle moves in a roughly circular, vertical orbit; the radius of the orbit is equal to about half the wave height. During the passing of the wave, the water particle follows a circular path and returns to its original position after the wave has passed. The deeper the water particle is from the surface of the ocean, the smaller is its orbit. Water particles at depths greater than half the wavelength have essentially no motion generated by surface waves. Waves in the open sea are called **waves of oscillation** because of this orbital motion.

The circular orbits of water molecules are flattened into oval patterns as the wave approaches the shallow water near the shore. Friction with the bottom begins to slow the wave down, and the upward slope of the bottom pushes the water upward to form higher waves. A high wave in which the crest falls forward in front of the main body of the wave is called a **breaker.** At this point, the waves have become **waves of translation.** As the water crashes onto shore, its motion is controlled by the back and forth energies in the surf zone. The still-turbulent sheet of water that sweeps up the slope of the beach is called the **swash;** the lower-energy water that flows back down the beach into the surf zone is called the **backwash.**

Waves generally approach the shore at an angle. The end of the angled wave closest to shore reacts to the decreasing depth by slowing, while the other end of the wave continues at full speed. Consequently, as the depth decreases, the wave crest bends to become more parallel with the shore. This process is called **wave refraction.**

Longshore currents and rip currents. Wave action continuously moves sand across or along the beach in the surf zone. Even after refraction most waves are still not exactly parallel to the shore. The push of these waves creates **longshore currents,** which carry sand parallel to the coastline and roll pebbles and gravel along the bottom. Longshore currents are usually quite strong and transport most of the sand in the shoreline environment. Generally, 1 to 2 million tons of sand are moved along a single beach environment every year.

A **rip current** is water that flows straight back out to sea after its waves have broken on the beach. These currents are most prominent immediately after a large set of waves has broken and tend to develop where wave heights are lower. They flow quickly back through the surf zone and dissipate in the open ocean. Rip currents look like fingers of discolored, muddy water that extend through the surf zone. Being caught in a rip current can frighten even the most experienced swimmers. Because the currents tend to be narrow, a person can swim out of one by swimming parallel to the shore *across* the current, not toward the shore against it.

Beaches

Beach features. The area of sand or gravel (more rarely silt) that covers the shoreline from the low-water edge to a well-defined upper elevation, such as a bluff or vegetated surface, is called a **beach.** The side of the beach facing the ocean is the **beach face** and is steepest because it experiences the greatest amount of erosion by wave action. Offshore beyond the beach face is the gently sloping **marine terrace,** a platform that may occasionally be seen at low tide. It is composed of sediment deposited by retreating waves or may be a bedrock surface that has been eroded by the waves. The landward edge of the beach is marked by the **berm,** the edge of a platform of sediment deposited by higher waves during severe storms. Berms can be quite coarse grained and contain abundant shell debris.

Beach deposition. The movement of sediment parallel to the shore by wave action is called **longshore drift.** A wave that washes across a beach face at an angle carries sand at that angle until it has lost all its energy; at that point, the water returns to the sea by running straight down the face of the beach into the surf zone. This process constantly moves sand across the beach face. The sand is carried the same way by the next wave, and moves across the face in a series of arcs. Called **beach drift,** this zig-zag pattern can transport sand and pebbles hundreds of meters a day along the beach.

The majority of the sediment in the beach environment is carried by the longshore current in the surf zone. The friction and erosion between the breaking waves and the bottom loosens and suspends sediment particles, which are then transported long distances in the current. The sediment is eventually deposited as fingerlike features called **spits** or **baymouth bars,** which can block an open bay from the ocean. A **tombolo,** a bar of sediment that connects an island to the mainland, forming a small peninsula, can also be formed.

Beach composition. Most beaches are composed of quartz sand, the majority of which is river sediment deposited in the ocean and reworked by ocean currents. Because of their high densities, black streaks and layers of heavy metallic minerals (magnetite, ilmenite) are also concentrated on beaches. Beach materials can also be limestone or basalt grains.

Seasonal changes. Wave action in the summer months tends to bring up sand from deeper water and builds wider beaches. Winter wave action generated by stormier weather erodes the sand and reduces the width of the beach. The sediment is carried out to sea and deposited as an underwater sandbar, which is then eroded by the next summer's waves to rebuild the beach.

Beach engineering. Engineering efforts designed to protect beaches and harbors interfere with sand drift and the natural development of beaches and coastlines. **Breakwaters,** built parallel to the shoreline to provide quiet water for pleasure boating, can result in extreme sedimentation that actually closes off the area. **Jetties** are walls that are built on both sides of a harbor and extend into the ocean to protect the harbor from excessive sedimentation and destructive waves. In most cases one jetty will trap the sand, resulting in "starving" the beach behind the other jetty, which begins to recede from erosion. **Groins** are series of walls built perpendicular to the coast to widen beaches that are losing sand to longshore drift. The natural beach environment represents an equilibrium between sand, wind, and waves. Human attempts to modify beach dynamics result in sedimentation patterns that generally frustrate the designers.

Shoreline Features

The **coast** is the strip of land near the ocean that includes the beach and the immediate inland area beside it. Coasts can be rocky and rugged or gently sloped. **Paleocoasts** are generally older coasts that were submerged under later marine transgressions and then lifted tectonically above sea level, exposing their sea floor features.

The constant impact of waves can dramatically alter even the most rugged, rocky coastline. Soluble rocks like limestones are dissolved, softer rocks are easily eroded, and even harder rocks like granite are fractured by the impact of waves. An irregular coast has many coves separated by irregular rocky points called **headlands.**

Wave action eventually straightens and smoothes the coastline. The headlands receive the greatest force of the waves, and the bays are the most sheltered. The headlands are broken down more quickly than the bays, and the eroded material is deposited in the coves. This process of the headlands being cut back and the flanking beaches being widened is called **coastal straightening** (Figure 38).

Coastal Straightening

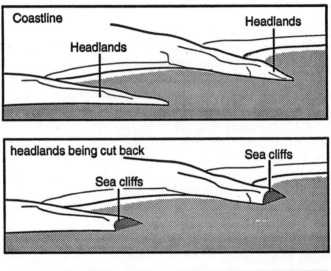

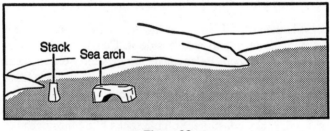

■ Figure 38 ■

Constant wave action along a rocky shore creates prominent **sea cliffs.** Sea cliffs retreat by mass wasting as waves undercut the cliffs in the wave zone, creating rock falls into the sea. **Sea caves** are cavities that are eroded into rock in the wave zone. As sea cliffs retreat, they leave behind, beneath the surf, a flat-lying bench of eroded rock called the **wave-cut platform.** **Stacks** are erosional remnants of sea cliffs that are rooted to the wave-cut platform and stand above the surface of the water. **Sea arches** are stacks whose centers have been eroded through because the rock is softer or more fractured, resulting in a bridgelike shape.

Depositional coasts are those gently sloped coasts that have been built up by sediments deposited from longshore drift. **Barrier islands** are large, elongate masses of sand that parallel the coast and form islands. These islands are separated from the coast by protected lagoons. Tidal currents may deposit **tidal deltas** in gaps between the barrier islands. Depositional coasts are also built outward by sedimentation in river deltas. Glacial sediments can also contribute to the growth of a coastline.

Biological activity is also important in stabilizing a coast. The development of offshore coral and algal **reefs** helps protect the coastline from erosion by being a barrier to strong wave action. Heavily vegetated coastlines with dense root systems, such as those of mangrove trees in Florida, anchor the beach, reduce erosion, and help trap sediments.

Estuaries are parts of old river channels that now extend inland from the coast. The shallow water in estuaries is typically brackish from the mixing of ocean water and fresh water.

Uplifted coasts are former coasts (paleocoasts) that have been lifted above the present coastline by tectonic activity. They are often identified by **uplifted marine terraces** that were formed below the older surf zone and are generally found along tectonically unstable coastlines, such as the Pacific coast in the United States and Canada.

Tides

The rhythmic rise and fall of sea level along a coastline is called the **tide.** The tide is a result of the gravitational attraction exerted upon the earth by the moon, and to a lesser extent by the sun. Tides occur about fifty minutes later each successive day for about twenty-nine days, which completes one cycle. At the times of the new and full moons, the earth, sun, and moon are aligned, causing the greatest difference in tidal elevations, called **spring tides. Neap tides,** producing the least extreme tidal differences, occur midway between the spring tides.

A **tidal current** is the horizontal flow of water that accompanies the tides and flows in one of two opposite directions. Tidal currents preceding high tide are called **flood currents;** tidal currents preceding low tide are called **ebb currents.** The zone of coastline affected by the tidal currents is called the **tidal flat.** Tidal action often generates cross-bedded marine sedimentary deposits called **tidal deltas.**

Areas that receive less than 25 centimeters (10 inches) of rain annually are called **deserts.** Deserts are dry with sparse vegetation. Landforms tend to have angular features because the lack of rain results in minimal chemical weathering, and flash floods create steep-walled scarps and gullies. There are few plants to protect the soil from the wind, so the soil is blown away to expose the rocky surface. Even in such a dry climate, most of the landforms are carved by the rare periods of heavy rainfall that result in flash floods, erosion, and sediment deposition.

Distribution and Causes of Deserts

Hot air rises at the equator, where the land receives the greatest amount of the sun's radiation. Most of the world's deserts are located near 30 degrees north latitude and 30 degrees south latitude, where the heated equatorial air begins to descend. The descending air is dense and begins to warm again, evaporating large amounts of water from the land surface. The resulting climate is very dry.

Other deserts are located in the **rain shadows** of mountain ranges. As moist air passes over a mountain range, it expands and cools, precipitating most of its moisture as it rises. As it sweeps down the other side of the mountain range, it warms and compresses, causing high evaporation rates and shedding little rain. Many of the deserts in the southwestern United States are the result of rain shadows.

A few deserts, such as the Gobi Desert in China, are simply a result of being located far from the ocean, from which most atmospheric moisture is drawn. The moisture is precipitated before it can reach these interior areas.

Deserts can form even on tropical coasts beside cold ocean currents, such as the west coast of South America. The currents cool the air, which then rises and warms as it moves over land, drawing up moisture that is later precipitated as the air moves farther inland.

Desert Features

Streams. Because of the dry conditions, most deserts do not have streams or rivers that run all year long. Streams that flow intermittently as a result of periods of sudden rainfall are called **ephemeral streams.** Exceptions are the Colorado River in the southwestern United States and the Nile River in Egypt, which originate in mountainous regions and have enough stream flow to cross desert areas.

Deserts often exhibit an **interior drainage** pattern where streams empty into landlocked basins. The basins become temporary sources of water, and evaporation can precipitate salt beds and other evaporitic minerals.

Flash floods. Most desert rainfall comes from short, violent thunderstorms. The rain is so abundant that it cannot soak into the hard ground or be controlled by the narrow stream channels. Rain (as sheetwash) flows rapidly over the land, creating **flash floods** in the stream beds that can be very destructive in populated areas. The lack of vegetation allows severe erosion, which carves new scarps and gullies; the water can become so choked with sediment it becomes a mudflow. The rapid downcutting by floodwaters produces narrow gorges with steep walls and gravel bottoms called **arroyos** or **dry washes** (**wadis** in Arabia and North Africa, **dongas** in South America, and **nullahs** in India).

Basin and Range topography. Some deserts, such as those in the American southwest, display **Basin and Range topography**—a series of steep mountain ranges separated by broad valleys. The mountain ranges cause a series of rain shadows that create the desert climate. Most of the rainfall in the mountains carries rock debris and sediment out to the alluvial fan that forms at the mouth of a canyon. If the water runs out farther into the center of the valley, it forms a **playa lake.** Typically shallow and muddy from clay, it evaporates quickly, leaving a flat, hard, dried clay surface that is broken by desiccation cracks. If the water carries dissolved salts, **salt flats** will result.

Eventually alluvial fans at the front of a mountain range may join to form a **bajada,** or rolling surface of sediment and gravel. Between the bajada and the range front is the **pediment,** a low-angle erosion surface at the foot of the mountain range that is typically covered by up to 30 meters (100 feet) of sediment. Isolated bedrock remnants of the former mountain front called **inselbergs** may project abruptly through the pediment cover as rocky hills.

Plateaus, mesas, and buttes. Hills underlain by resistant rock such as sandstone, limestone, or volcanic lava are called **plateaus.** Plateaus are edged by steep-sided scarps and gullies. As weathering and erosion cut back a plateau's slopes, remnant flat-topped towers or columns called **mesas** may be left behind. The continued erosion of a mesa results in a similar but narrower landform called a **butte.** (The plateau-mesa-butte sequence is an example of parallel retreat.) Although most common in desert climates, these landforms are more a function of rock structure than climate.

The Effects of Wind

Bed load and suspended load. Winds in the desert are often extreme and unrestricted by trees and vegetation. Wind can be an effective erosion and transportation agent if it is strong and blows across fine-grained sediment such as sand, silt, and clay. A wind's **bed load** consists of the heavier grains (usually sand) that hop and skip along the ground by saltation. These rarely rise more than 1 meter (3 feet) into the air as they are transported. The **suspended load** is the finer-grained clay and silt fraction that is actually carried long distances in the wind.

Wind velocity. The velocity of wind is a result of air pressure differences due to heating and cooling. Desert winds are the result of temperatures that commonly fluctuate from 7 degrees centigrade at night to 43 degrees centigrade or more during the day (45–110 degrees Fahrenheit) and can travel at speeds up to 70 miles per hour.

Dust storms. Depending on the amount of fine-grained material that is available and the speed of the wind, **dust storms** that blot out the sun can result. Particles can be carried thousands of feet upward into the atmosphere and for hundreds of miles laterally. Dust storms stripped the fertile soil from the overfarmed and drought-stricken fields of the "Dust Bowl" in the United States in the 1930s. A small but regular component of land-derived sediment is deposited in the ocean. Volcanic ash from famous eruptions such as Krakatoa was carried around the world by winds for two years. Abrasive windblown sand carves rocks and boulders into unusual shapes called **ventifacts,** which have flat, wind-abraded surfaces.

Deflation. **Deflation** is the removal of sediment from a land surface by wind. It can lower the surface of land significantly by forming a bowl-like depression called a **blowout.** Blowouts can be over 60 kilometers in diameter and over 40 meters deep. Another result of deflation is thought to be **desert pavement,** a large surface of the desert floor that is covered by pebbles and stones that resemble rounded paving stones. Some geologists believe that the wind removes the fine-grained material from the surface until only coarser material remains; others suggest that the pebbles move up through the fine-grained sediment by thermal expansion and contraction (much like frost heaving). Desert pavement is likely the result of a combination of both processes.

Silt and clay that is deposited by wind is called **loess.** Typically very porous, it forms downwind from the source and blankets hills or accumulates in depressions. It is typically cemented by calcite. Loess can reach thicknesses of 100 meters (300 feet). The fertile soils of the midwestern and Pacific northwestern United States include loess.

Sand dunes. Sinuous heaps of unconsolidated sand called **sand dunes** are the classic feature of the great deserts of the world. Dunes are deposited by winds in desert regions or along sandy coastlines and beaches. Dune material varies in composition and includes sand-size grains of quartz, feldspar, calcite, gypsum, and rock fragments that are well sorted and well rounded.

A dune's shape is constantly changing according to the wind direction. The steeper, downwind slope is called the **slip face.** The loose sand maintains angles up to about 35 degrees on the slip face, creating cross-bedded layers. Dunes migrate in the direction of the prevailing winds about 12 meters (40 feet) per year, a result of the wind continually eroding the gentle slope and redepositing the sand on the slip face. The surface is typically marked by a series of sand ripples.

Figure 39 illustrates various sand dunes in planar (flat) view from above. One of the largest and most dramatic dunes is the **longitudinal dune, or seif.** A large ridge of sand that parallels the wind direction, it can be over 100 meters high and over 100 kilometers long. **Barchan dunes** are widely separated, crescent-shaped dunes that form in areas of sparse sand. Often found on bedrock, the ends of the crescents point downwind. **Transverse dunes** are a series of long ridges that form perpendicular to the wind. They typically occur in coastal areas. A **barchanoid dune,** an intermediate variety between barchan and transverse dunes, is formed of scalloped rows of sand perpendicular to the wind. It resembles a series of side-by-side barchan dunes. A **parabolic dune** usually forms around a blowout in vegetated areas—the dune is deeply curved and the tips point into the wind. **Star dunes** are isolated hills of sand formed by variable winds in the Sahara and Arabian Deserts. The bases of these dunes resemble multipointed stars.

Types of Dunes

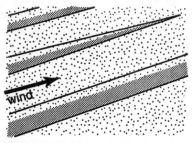

Planar view of longitudinal dunes

Planar view of barchan dunes

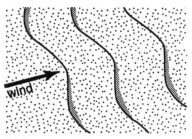

Planar view of transverse dunes

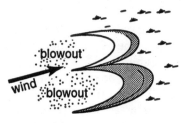

Planar view of parabolic dunes

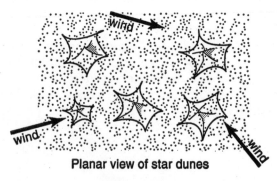

Planar view of star dunes

■ Figure 39 ■

Depending on how strong they are and where they strike, earthquakes can be some of the worst natural disasters, taking thousands of lives and creating billions of dollars of damage. Earthquakes are the result of the sudden movement of rock along a fault zone beneath the surface, usually centered in tectonically active areas. An earthquake beneath the ocean results in huge waves called seismic sea waves (tsunamis) that devastate coastlines. Scientists who study earthquakes (**seismologists**) hope to eventually predict earthquakes before they strike. Engineers are also challenged to develop new designs that will enable buildings to withstand the forces of earthquakes.

How Earthquakes Form

An **earthquake** is the shaking of the earth caused by the sudden release of energy from rocks under tectonic stress. Most earthquakes are associated with rock movements along faults below the surface of the earth. Because of friction and high confining pressure the fault blocks don't move until the tectonic stress becomes great enough to overcome the frictional force. Earthquakes may also result according to the **elastic rebound theory,** which suggests that in some cases energy is *stored* in rock that is being bent (deformed) by tectonic forces until the energy in the rock exceeds the bonding strengths between minerals, and the rock breaks. When the rock breaks, it suddenly returns to its predeformed shape, and the crust moves violently as a result of the quickly released force. This results in the formation of a new fault. As an analogy, you might think of a flexible plastic rod being bent until it reaches its breaking point. After it breaks, its two ends spring back from their curve into a straight line, the energy that curved them having been released. The displacement along the fault zone typically ranges from about 1 to 7 meters; earthquakes rarely last longer than a few minutes.

Not all earthquakes are associated with existing faults—some are related to deeply buried fold structures, thrust faults, and volcanic environments in which magma is forcing its way to the surface. Earthquakes can also create new faults. Some geologists also theorize that the deeper earthquakes are a result of mineral transformations in cold subducted plates that enter the hotter mantle.

Seismic Waves

When rock masses suddenly move deep within the earth in response to tectonic stress, energy in the form of **seismic waves** moves outward through the rock from the point of origin, called the **focus**. The initial movement occurs at the focus. The **epicenter** is the point on the surface directly above the focus. There are three types of seismic waves: P and S body waves and surface waves.

Body waves. **Body waves** radiate outward from the focus in all directions and travel through solid rock. A **P body wave** (**primary body wave**) is a compressional (longitudinal) wave that induces the particles in the rock to vibrate back and forth in the same direction the wave moves. P waves move at speeds up to 15,000 miles per hour (7 kilometers/second) (Figure 40).

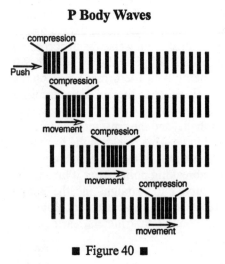

P Body Waves

■ Figure 40 ■

An **S body wave (secondary body wave)** is only about half as fast as a P wave and causes the rock to vibrate at right angles to the direction of wave travel (Figure 41).

S Body Waves

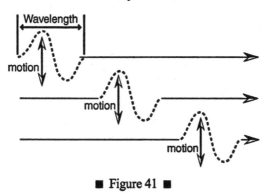

■ Figure 41 ■

Both P and S waves can travel through solid rock, but only P waves can pass through a fluid medium.

Surface waves. **Surface waves** are the slowest seismic waves and travel outward on the earth's surface from the epicenter much like ripples do from a stone thrown into water (Figure 42). They create most of the damage at the surface because they are the waves that produce the most ground movement and pass through an area the most slowly.

Surface Waves

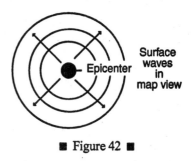

■ Figure 42 ■

Monitoring Earthquakes

Seismometers. Seismic waves are detected with a **seismometer,** which contains a suspended pendulumlike mass that is kept as motionless as possible. The seismometer is connected to a **seismograph,** which records the motion of the mass during an earthquake as a series of squiggly lines called a **seismogram.** Seismograph stations all over the world can record the seismic waves from the same earthquake. The location, depth, and strength of an earthquake can easily be calculated from the seismograph data. When the data are compared from station to station, they can also reveal clues about the nature of the rocks through which the waves passed.

Depth of focus. Seismic wave data can also be used to calculate the **depth of focus,** or the vertical distance between the epicenter and the focus. The maximum depth for earthquakes is about 670 kilometers (400 miles). Eighty-five percent of all earthquakes have a shallow focus that can range as deep as 70 kilometers (40 miles); 12 percent have an intermediate focus that ranges from 70 to 350 kilometers (40–210 miles); the final 3 percent (deep focus) originate at depths of 350 to 670 kilometers (210–400 miles). Most earthquakes have a shallow focus because it is in this area that rocks are brittle and can break more easily—at greater depths the rocks flow more plastically in response to tectonic stress.

Travel-time curves. P and S waves originate from the focus simultaneously. Because they travel at different speeds, they arrive at various seismograph stations at different times. The farther the station is from the focus, the farther apart are the waves when they arrive. This time interval is used to plot a **travel-time curve,** which can determine how far a station is from the origin of the earthquake. The station then plots a circle on a map with its computed distance as the radius of the circle. The location of the earthquake is the point at which three or more circles from different stations intersect on a globe (Figure 43).

Locating Earthquakes

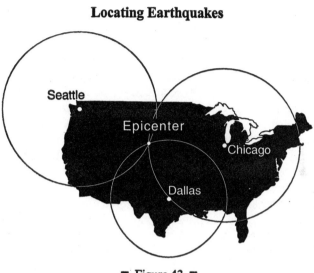

■ Figure 43 ■

First-motion studies. Seismic data can also be used to determine in which direction rocks first moved along a fault during an earthquake. These **first-motion studies** indicate if the first rock motion was a push (the rock moved toward the seismograph station) or a pull (the rock moved away from the station). An analysis of push-pull data can generate two possible fault orientations that fit the first-motion data. If the fault orientation is known, the direction of displacement along the fault can be accurately selected.

Intensity. The strength of an earthquake can be measured as a function of **intensity.** The **modified Mercalli scale** ranks intensity from 1 to 12 according to the amount of resulting damage. This system is not totally accurate because the amount of damage is often proportional to the population in an area, the type of design and construction of buildings, and the base on which the buildings sit (that is, bedrock or sediment). Another limitation is that intensities cannot be assigned to earthquakes in uninhabited areas because there is little physical damage that can be quantified.

Magnitude. The most common system for quantifying the strength of an earthquake is by its **magnitude.** By analyzing the seismic waves, the magnitude, or the amount of energy released by the earthquake, can be determined. The **Richter scale** is a numerical scale that lists earthquake magnitudes in logarithmic increments from about 2 to 8.6—the highest value ever recorded on the scale. The logarithmic relationship means that the difference in the *amplitude* of the vibration in a seismic wave between two consecutive whole numbers on the scale is a factor of ten. Thus, an earthquake that is a 3 on the Richter Scale has a vibration ten times bigger than that for a 2. The difference in energy released is even larger: a Richter 3 is about thirty times more powerful than a Richter 2.

The Richter scale is fairly accurate up to about 7. For more powerful earthquakes, the magnitude is now being calculated according to the amount of resulting surface rupture and fault displacement. To date, this new method of measurement has calculated a maximum value of 9.5 (the 1960 earthquake in Chile).

On a yearly basis, more than 100,000 Richter 2 ("just felt") earthquakes are recorded, several thousand Richter 4 to 5 earthquakes result in significant damage, and about 15 to 20 earthquakes classified as 7 or greater on the Richter scale cause serious damage. The majority of earthquakes in the United States occur in tectonically active areas, such as along the Pacific coast and fault zones in the western states. Although only a few earthquakes occur in the central and eastern states, they can be quite damaging because the crust is older and more brittle.

Effects of Earthquakes

Structural damage and fire. Surface trembling from seismic waves often damages buildings. Depending on the severity of the earthquake, gas mains may break, starting numerous fires. **Foreshocks,** small earthquakes that sometimes precede the main earthquake, can be used as a warning system that a large shock may be on the way. Thousands of **aftershocks** may follow an earthquake and can be quite destructive, especially to those structures that have already been weakened and damaged.

Mass-wasting events. Ground motion may trigger landslides and other rapid mass-wasting events that result in loss of life and damage to buildings. A mass-wasting variation is a landslide by **liquefaction,** in which water-soaked sediment moves downslope like a slurry. Buildings that were built on solid sediment may sink if liquefaction occurs.

Rocks can be permanently displaced during an earthquake. Fault blocks may move vertically, forming a new scarp along the fault plane. Horizontal movement can tear apart roads, pipelines, and any other structures that are built across the fault zone. Displacement rarely exceeds about 7 meters (25 feet).

Seismic sea waves. If the sea floor suddenly shifts upward or downward, the sudden displacement of water results in **seismic sea waves,** or **tsunamis.** Unlike even the greatest storm waves, tsunamis can be up to 90 meters (300 feet) high and move at speeds of up to 400 miles per hour. Tsunamis have wavelengths that can be as long as 160 kilometers (100 miles), and the water does not quickly withdraw from the coast after the tsunami breaks. The water continues to rise for up to ten minutes until the long wavelength has passed through, resulting in widespread coastal damage.

Earthquakes and Plate Tectonics

Earthquake belts and distribution. Earthquakes occur in well-defined belts that correspond to active plate tectonic zones. The **circum-Pacific belt** (also called the **Rim of Fire**) follows the rim of the Pacific Ocean and hosts over 80 percent of the world's shallow and medium-depth earthquakes and 100 percent of the deep earthquakes. Other earthquake zones are the **Mediterranean-Himalayan**

belt and the **midoceanic ridges** that split the crust at the bottom of the world's oceans.

Plate boundaries and associated earthquakes. Distribution plots reveal that many earthquakes are associated with andesitic volcanic action and oceanic trenches that occur over subduction zones in the circum-Pacific belt. **Oceanic trenches** are narrow, deep troughs that mark where two plates converge, usually along the edge of a continent or island arc where andesitic volcanoes typically occur. Earthquakes originate in **Benioff zones,** zones that slope downward from the trenches and under the overlying rocks at 30 to 60 degrees. Benioff zones are closely associated with the subduction of a crustal plate below an adjacent plate.

Almost all earthquakes occur at the edges of the crustal plates. The constant bumping, grinding, and lateral movement along crustal boundaries can create sudden movements that result in earthquakes. Each of the three types of plate boundaries—convergent, divergent, and transform—has a distinctive pattern of earthquakes.

There are two kinds of **convergent boundaries:** subduction and collision. A **subduction boundary** is marked by the oceanic crust of one plate that is being pushed downward beneath the continental or oceanic crust of another plate. A **collision boundary** separates two continental plates that are pushed into contact; the **suture zone** is the line of collision. Both types of boundaries have distinctive earthquake patterns.

Earthquakes associated with a collision boundary define shallow, broad zones of seismic activity that form in complex fault systems along the suture zone. Earthquake patterns in subduction zones are more complex. As the oceanic crust begins to descend, it begins to break into blocks because of tension stress. Shallow earthquakes in the upper part of the subduction zone are a result of shallow-angle thrust faults, in which slices of plates slide like cards in a deck that is being shuffled. Earthquakes also periodically occur as the plate continues to subduct up to a depth of about 670 kilometers (400 miles).

First-motion studies of these earthquakes suggest they result from both compressional and tensional forces on the subducting plate.

Earthquakes are relatively abundant in the first 300 kilometers (180 miles) of a subduction zone, are scarce from 300 to 450 kilometers (180 to 270 miles), and then increase slightly again from 450 to 670 kilometers (270 to 400 miles). It is possible that these deepest quakes are related to sudden mineral transformations and resultant energy releases or volume changes. It has been theorized that earthquakes do not occur at depths greater than 670 kilometers because the subducting plate is not brittle anymore and has become hot enough to flow plastically.

The distribution of earthquake foci along a subduction zone gives an accurate profile of the angle of the descending plate. Most often, plates start subducting at a shallow angle, which becomes steeper with depth. The angle of subduction is proportional to the density of the plate material, the amount of faulting and thrusting, and the tearing or crumpling of the descending plate.

Divergent boundaries are those at which crustal plates move away from each other, such as at midoceanic ridges. These huge underwater mountains often have a central graben feature, or **rift valley,** that forms at the crest of the ridge. The formation of new ocean crust that is pushed away from both sides of the ridge fault creates a tensional setting that results in the formation of the graben. Earthquakes are located along the normal faults that form the sides of the rift or beneath the floor of the rift. Divergent faults and rift valleys within a continental mass also host shallow-focus earthquakes.

Shallow-focus earthquakes occur along **transform boundaries** where two plates move past each other. The earthquakes originate in the transform fault, or in parallel strike-slip faults, probably when a frictional resistance in the fault system is overcome and the plates suddenly move.

Control and Prediction

Monitoring. There is great interest in learning enough about earthquakes to be able to predict their occurrence and possibly even control them. Rock adjacent to a fault in a seismically active area can be monitored for slight changes, such as the development of small cracks, very small tremors called **microseisms,** and changes in magnetism, electrical properties, and seismic velocity. The formation of cracks in the rock increases the rock's porosity, which can change the local water table—a phenomenon that has been noticed before earthquakes. Similarly, the cracks allow radioactive gases like radon to escape, which can also be detected. Minute changes in the elevation of the earth's surface can also be measured. Asian scientists have documented that various species, including insects, horses, and snakes, are ultrasensitive to seismic activity and can become skittish before an earthquake.

All of these changes are thought to reflect increasing strain that may indicate the onset of an earthquake; however, such changes are not foolproof indicators and may apply to one earthquake and not another. It is also possible that fault zones may actually be quite weak and not require great amounts of strain to create movement.

Using historical data. The probability of an earthquake striking an area can be calculated based on the historical seismic activity in that area. Repeated patterns of earthquakes make it possible to predict when the next one *might* occur. **Seismic gaps** are stretches along an active fault zone that have not produced earthquakes for a significant time. Scientists believe that these gaps may be extra-resistive areas along the fault where strain is being stored in the fault blocks and where earthquakes will likely occur when the strain is finally released.

Reducing built-up strain. Damage from earthquakes in heavily populated areas could be reduced if the built-up strain could somehow be released periodically before the rock breaks or moves. It has been accidentally discovered that pumping water under high pressure into the subsurface can trigger earthquakes. The water probably lubricates the fault zones, causing movement, or helps dissipate the strain that was being stored in the rocks. A similar technique might be developed that could release the enormous accumulation of rock strain in quake-prone areas through a series of nondestructive, low-magnitude, timed quakes.

The rocks that can be studied on the earth's surface tell us much about the earth's uppermost crust but very little about the other 99 percent of the planet. Drilling has reached a maximum depth of about 12 kilometers (7 miles). Samples of rock from the deeper crust and mantle are sometimes included as xenoliths in deep-seated intrusives that moved along structural zones to the surface. Rare segments of ultramafic rocks in complex tectonic settings are thought to have originated from the lower crust or upper mantle.

Fortunately, the field of **geophysics**—the application of the laws of physics to the dynamics of the earth—provides compelling data that allow us to interpret how the inner earth is constructed. The principal characteristics that geophysicists study are seismic waves, gravity, heat flow, magnetism, and electrical conductivity. When integrated, the data allow us to construct a realistic picture of how the inner earth works.

Methods of Detection

Because seismic waves from earthquakes (or surface explosions) can pass through the entire earth, the behavior of these waves permits deductions about the rocks through which they have passed.

Seismic refraction. The direction of travel of a seismic wave, like that of a beam of light, can be bent, or refracted, when it passes into or out of different mediums. This **seismic refraction** occurs only if the mediums have different densities or strengths, which change the velocity of the seismic wave (Figure 44).

Seismic Refraction

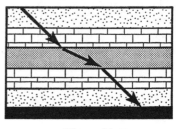

■ Figure 44 ■

A distinct rock boundary is not necessary to bend a seismic wave. Seismic waves begin to develop a slightly curved path as they move away from the source, a result of many small refractions as the waves pass through the different rock types of the crust.

Seismic reflection. **Seismic reflection** is the return of some of the energy from seismic waves that have penetrated downward from the surface or near-surface, hit a rock boundary, and bounced back to the surface (Figure 45). Since the time of departure and return are known from the seismogram, the depth to the rock boundary can be calculated.

Seismic Reflection

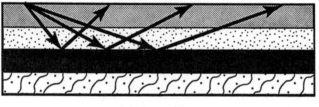

■ Figure 45 ■

The Structure of the Earth

By analyzing seismic refraction and seismic reflection data from all over the world, geophysicists have identified boundaries that separate three concentric parts of the earth: the crust, mantle, and core. The **crust**—5 to 50 kilometers (3–30 miles) thick—is essentially the thin, hardened skin of what was once the earth's molten exterior surface. Underlying the crust is the **mantle**—2,900 kilometers (1,740 miles) thick—a thick zone of much hotter, partially plastic rock. The mantle surrounds the inner and outer **core**—about 7,000 kilometers (4,200 miles) combined in diameter—the innermost zone of the planet. (See Figure 1, p. 2.)

The Crust

Crust composition. P waves travel faster in oceanic crust (7 kilometers/second) than in continental crust (6 kilometers/second)—these speeds are about the same as those through basalt/gabbro and granite/gneiss, respectively. This suggests that oceanic crust is mostly basaltic and that continental crust is mostly **sialic,** meaning the rocks, such as granite, contain high amounts of aluminum and silica. Oceanic crustal rocks, such as gabbros and basalts, are high in magnesium and silica (sometimes called **sima).**

Crust thickness and density. The seismic data also show that the thickness of the continental crust ranges from 30 to 50 kilometers (18–30 miles) and that of the oceanic crust from about 5 to 8 kilometers (3–5 miles). Continental crust is thickest under mountain ranges, where it bulges downward into the mantle, forming a **mountain root.** Geophysical data also show that continental crust would "float" on oceanic crust because continental crust is less dense (continental crust, 2.7 g/cm³; oceanic crust, 3.0 g/cm³).

The Mohorovicic discontinuity. The **Mohorovicic discontinuity,** or **Moho,** the first major boundary of the earth's interior, separates the crust from the underlying mantle. It is named for Yugoslavian seismologist Andrija Mohorovicic, who in 1909 presented the first evidence for the layered internal structure of the earth. The Moho occurs at a depth that ranges from 5 to 50 kilometers (3–30 miles) from the surface.

The Mantle

The upper and lower mantle. Seismic data suggest that most of the mantle is composed of solid rock. P waves travel at an average of about

8 kilometers per second through the mantle, suggesting it is composed of ultramafic rocks such as peridotite. The behavior of P waves indicates the mantle can be divided into two parts: the upper and lower mantle. The **upper mantle** begins at a depth of from 5 to 50 kilometers (3–30 miles) and extends to a depth of approximately 670 kilometers (400 miles) from the surface; the **lower mantle** extends from a depth of about 670 kilometers (400 miles) to about 2,900 kilometers (1,740 miles).

The lithosphere. Changes in P wave velocities have identified other boundaries in the mantle. Together the crust and the uppermost mantle form the **lithosphere.** This brittle exterior shell of the earth ranges in thickness from about 75 kilometers beneath oceans to about 175 kilometers beneath continental masses (45–105 miles). The maximum depth of the lithosphere from the surface is thought to be no more than 200 kilometers (120 miles).

The asthenosphere. P waves slow down at the boundary between the lithosphere and the underlying **asthenosphere.** Most geologists believe the asthenosphere, also called the **low-velocity zone,** is about 200 kilometers (120 miles) thick. The rocks in the asthenosphere are thought to be partially melted and hotter and more plastic than those in the lithosphere. If this is the case, the asthenosphere could be the weaker surface on which the crustal plates move and a possible source for the generation of magma.

Other boundaries. Seismic data has identified two other concentric boundaries within the upper mantle at depths of 400 and 670 kilometers (240 and 400 miles). These are thought to represent pressure-temperature zones in which minerals collapse to form new, more compact atomic structures. It is therefore possible that even though the mantle may be chemically homogenous, it does not have the same mineral composition throughout. The 670-kilometer boundary is believed to represent a chemical as well as physical boundary between the upper and lower mantle. The thick lower mantle is believed to consist of plastic or partially melted ultramafic rocks that continue to the boundary with the outer core.

Isostatic Equilibrium

Huge plates of crustal and upper mantle material (lithosphere) "float" on more dense, plastically flowing rocks of the asthenosphere. The "depth" to which a plate, or block of crust, sinks is a function of its weight and varies as the weight changes. This equilibrium, or balance, between blocks of crust and the underlying mantle is called **isostasy.** The taller a block of crust is (such as a mountainous region), the deeper it penetrates into the mantle because of its greater mass and weight. Isostasy occurs when each block settles into an equilibrium with the underlying mantle. Blocks of crust that are separated by faults will "settle" at different elevations according to their relative mass (Figure 46).

Isostasy

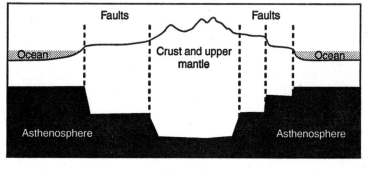

■ Figure 46 ■

The isostatic relationship is maintained as the crustal surface changes. For example, as a mountain range block erodes, the block will rise—it is not as heavy because the material is eroded away, and it does not need to "ride" as low in the mantle. The eroded material is deposited as sediment on the adjacent thinner continental blocks, which increases their weight, and they then sink farther into the plastic asthenosphere. Areas that are tectonically stable tend to be isostatically balanced. The viscosity of the mantle can be calculated based on the rates of the isostatic adjustment of the crustal blocks.

The development of thick ice sheets during the Pleistocene epoch warped the underlying crust downward into the mantle, an isostatic adjustment in response to the great weight of the ice. After the ice melted, the weight was removed from the crust and it began to slowly rise back to its preglacial position. This isostatic process, called **crustal rebound,** is still in progress in the Great Lakes area of the United States.

Some geologists believe that plate subduction generates large bodies of magma that adhere to the bottom of the continental mass and cool, locally thickening the crust. In order to maintain isostasy, the crust would then have to rise through the formation of a mountain range. This idea has not been widely accepted, however.

The Core

Shadow zones. The behavior of P and S seismic waves has been used to identify the presence of the core. When P waves originate from an earthquake and encounter the core, they are refracted inward. This refraction creates two areas on the opposite side of the earth where P waves are not detected. Called **P-wave shadow zones,** these are the intervals on the surface between the last unrefracted P wave and the first refracted P wave (Figure 47). Knowing how P waves behave allows the location and shape of the core to be estimated from P-wave data.

P-Wave Shadow Zone

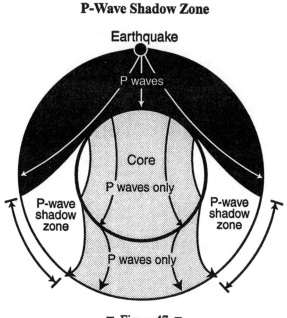

■ Figure 47 ■

S waves cannot penetrate the region of the core at all, creating an extensive **S-wave shadow zone** across about half of the earth's surface (Figure 48). Since S waves can pass through only solid material, it is very likely that at least the outermost core is liquid or molten.

S-Wave Shadow Zone

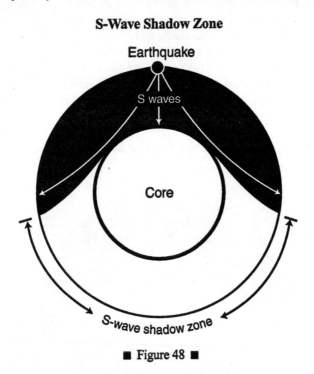

■ Figure 48 ■

Combining this information with a more detailed analysis of P waves traveling through the core, geophysicists think the core has two parts: a solid inner core and a liquid outer core.

Other data. A combination of seismic data and studies of the speed of the earth's rotation on its own axis and around the sun can also help determine the composition of the core. The average density of the earth is about 5.5 g/cm³; the density of crustal and mantle rocks varies from about 2.7 to 5.5 g/cm³. Since the crust and mantle combined make up over three-quarters of the earth, the core must have a density ranging from about 10 to 13 g/cm³. This kind of density can be achieved by having a composition of mostly iron mixed with small amounts of oxygen, sulfur, or silica. Some meteorite fragments that have been found approach this composition and are thought to represent primitive pieces of our solar system. The fact that the earth has a strong magnetic field also suggests its core is metallic.

A prominent transition boundary separates the core from the mantle. Up to 200 kilometers (120 miles) thick, the boundary is marked by different densities and temperatures. It may represent the lower limit of mantle convection, where colder, higher-density material such as subducted plates are reassimilated into the mantle. It may also be the upper limit for convection in the core, where hotter, lower-density material rises from the center of the earth, cools, and sinks again.

Magnetic Fields

The magnetic field and the poles. The earth is surrounded by a **magnetic field.** Magnetic lines of force originate from north and south **magnetic poles,** which are about 11.5 degrees away from the geographic North and South Poles. The magnetic field is strongest at the magnetic poles. The positions of the magnetic poles have changed over time and appear to be rotating around the geographic poles on an axis tilted from the geographic axis by 11.5 degrees. The magnetic field traps high-energy particles created by the sun's ultraviolet radiation, thus protecting our environment on Earth.

The magnetic field is thought to be generated by the liquid outer core. If this liquid material is metallic, as geophysical studies suggest, its flow as a result of heat convection would create an electric current. Electric currents induce magnetic fields.

Magnetic anomalies. The intensity of the magnetic field is measured at the earth's surface with a **magnetometer.** Large-scale patterns may be related to convection patterns in the liquid outer core. Local magnetic features or anomalies are usually related to different rock types. Rocks have different magnetic characteristics that, when added to the overall regional magnetic pattern, create anomalies. **Magnetic anomalies** are areas of magnetism that are either higher or lower than the average magnetic field for the area. A **positive magnetic anomaly** is a reading that exceeds the average magnetic field strength and is usually related to more strongly magnetic rocks, such as mafic rocks or magnetite-bearing rocks, underneath the magnetometer. A **negative magnetic anomaly** is a reading that is lower than the average magnetic field. Positive anomalies can also be created by irregularities in the bedrock surface beneath sedimentary cover; a rock that is only 10 meters from the surface and buried by sediment will have a more positive magnetic reading than the same rock that is 80 meters from the surface and covered by sediment. Similarly, negative anomalies can result from troughs or grabens that have developed on the bedrock surface.

The magnetic characteristics of the bedrock, especially in areas covered by glacial sediments, can be mapped in great detail using magnetic-field values. The magnetic data can even show the strike and dip of the rock units and outline the contacts between rock units of different magnetism.

Polarity reversals. The earth's magnetic field has periodically reversed its polarity in the geologic past: north becomes south, and south becomes north. This phenomenon is known from rocks that formed during these periods of reversal. Magnetic minerals crystallize in cooling lava flows and point themselves toward the north magnetic pole. This magnetic record is permanently trapped in the rocks when they harden. The study of **paleomagnetism** involves the identification of older magnetic fields that surrounded Earth in the geologic past.

The best source rocks for detailed paleomagnetic studies are thick accumulations of flood basalts in the interiors of continental plates. There have been over twenty paleomagnetic reversals in about the last 5 million years; our current magnetic field orientation has been stable for the last 700,000 years. An average for the past 20 million years is about one field reversal every 500,000 years.

It has been theorized that magnetic reversals are a result of changes in direction of convection flow in the liquid outer core or of periods of no convection. It is likely, before the onset of a paleomagnetic reversal, that there is a brief period of zero magnetic field, which may allow ultraviolet radiation to bombard the earth's surface and damage or kill various species; in fact, some species extinctions and mutations correlate with some paleomagnetic reversals.

Gravity

A **gravity meter** measures the force of gravity between a mass inside the instrument and the earth. The force of gravity between two objects varies according to the mass of the objects and the distance between them: if either mass increases, the force between them increases; if the distance between them increases, the gravitational force decreases.

Dense rocks attract the mass in the gravity meter more than do those rocks that are less dense. A rock has a **positive gravity reading,** or **anomaly,** if it has a value higher than the normal regional gravity value; it has a **negative gravity reading,** or **anomaly,** if it has less than the regional gravity value. For example, ultramafic rocks or metallic ore bodies are very dense rocks that give positive gravity readings.

If gravity measurements are the same across a series of crustal blocks, it indicates the region is in isostatic equilibrium. The reverse is true if the gravity measurements are variable. Positive and negative gravity anomalies are often the result of an area's being out of isostatic equilibrium due to tectonic forces that either hold a region up (mountain range) or hold it down (ocean trenches), respectively.

Gravity anomalies on a local scale are usually associated with changes in rock type and can be used as a mapping tool, especially in areas where the bedrock is hidden by sedimentary cover. Metallic ore bodies can also result in a positive gravity feature because the metallic minerals are much denser than the surrounding rock.

Geothermal Gradients

The internal temperature of the earth increases with depth from the surface. Near the surface, the average **geothermal gradient** is about 25 degrees centigrade (77 degrees Fahrenheit) for every kilometer of depth. Some areas have much higher heat flows because of deep fault zones, rifting, magmatic intrusions, or active tectonic forces. The geothermal gradient can make conditions in deep mines quite uncomfortable and hot enough to explode rocks or bend steel.

The geothermal gradient of 25 degrees centigrade/kilometer is thought to be restricted to the upper part of the crust. If it continued at this rate uniformly from the surface, the internal temperature of the earth would be greater than 2,000 degrees centigrade within the lithosphere—a temperature that far exceeds the melting temperatures for all rocks at that depth. Since the crust and upper mantle are solid and brittle, this gradient cannot extend to these depths, where it is more likely about 1 degree centigrade/kilometer. Recent laboratory studies have suggested the temperature is about 4,800 degrees centigrade at the base of the lower mantle and about 7,000 degrees centigrade in the inner core.

The **heat flow** is the amount of heat from the earth's interior that is lost at the surface. The heat is probably generated by a still-cooling core or by the radioactive decay of elements such as uranium and thorium. Areas of higher heat flow are generally related to magmatic activity or tectonic forces that bring wedges of hot mantle rock (mantle plumes) into thin or faulted crustal areas.

It is estimated that the world's first oceans formed about four billion years ago as a result of the cooling of the primitive earth. It is thought that volcanic eruptions discharged huge amounts of hot water vapor and other gases that cooled to form liquid water on the rugged surface. The volume of the oceans grew as volcanoes continued to emit gases from the molten rocks below. The chemical composition of the ocean water became salty as sodium, calcium, and magnesium were freed by chemical weathering and swept into the ocean by erosional processes. Chlorine and other elements were contributed from other volcanic gases.

Investigative Technologies

Oceans cover about 70 percent of the earth's surface. The geology of the sea floor was largely unknown until the last half of the twentieth century, when the rapid advance of new technologies allowed geologists to study the sea floor in great detail.

Mapping the topography of the sea floor is accomplished using an **echo sounder.** This procedure calculates the depth to the ocean floor using the time it takes a sound wave sent from a ship to bounce off the bottom and return. A **seismic profiler** is similar to an echo sounder but uses lower frequencies that penetrate farther and result in more detailed profiles that show underlying structures such as faults and buried topography.

Other sampling techniques include using a **rock dredge** to collect rock samples from the bottom and a **corer** to retrieve a column of sediment in a weighted steel pipe. Huge derricks allow drilling directly into the ocean floor to recover solid cylindrical cores of ocean bedrock. **Submersibles** such as the HMS *Challenger* have taken geologists down to the ocean floor faults and allowed them to directly observe the geologic features, including active hydrothermal vents precipitating mineral deposits.

Continental Margins

Active and passive margins. Continental margins are defined as active or passive according to the presence or absence, respectively, of plate tectonic activity. Earthquakes and volcanoes are associated with **active continental margins,** which are marked by a landward continental shelf, a much steeper continental slope that ends at an active ocean trench, and an irregular ocean bottom that may contain volcanic hills (Figure 49).

An Active Continental Margin

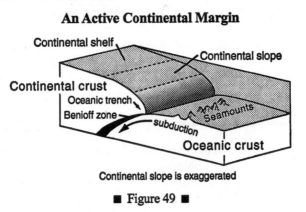

Continental slope is exaggerated

■ Figure 49 ■

A **passive continental margin** has a landward, shallow continental shelf, a deeper continental slope, a continental rise, and a flat abyssal plain (Figure 50).

A Passive Continental Margin

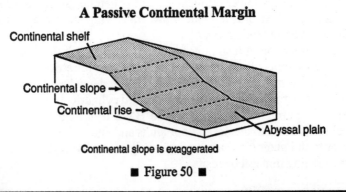

Continental slope is exaggerated

■ Figure 50 ■

Continental shelves. A **continental shelf** is a shallow, almost flat platform that extends seaward from the edge of the continent. The nearshore sediment is mostly sand that grades outward toward fine-grained mud at the deeper edge. Continental shelves range in width from a few kilometers to over 1,000 kilometers; depths increase from a few meters to about 200 meters. Being well within the 200-meter deviation for sea level variations during glacial epochs, sedimentation on continental shelves frequently shows marine transgressions and regressions. The continental shelf is underlain by sialic (high in silicon and aluminum) crust, which is part of the continental mass. Continental shelves cover about 8 percent of the ocean floor.

Continental slopes. The **continental slope** extends from the seaward edge of the continental shelf into the deep ocean (15,000 feet) at an average angle of 4 to 5 degrees. It is thought that the sediments of the continental slope cover the transition zone between continental and oceanic crust, a zone that may be structurally complex and contain block faults and thrust faults.

Submarine canyons. **Submarine canyons** are erosion features that cut continental shelves and slopes. The heads of some of these V-shaped canyons may have been carved by river erosion when the sea level was lower during Pleistocene glaciation. These canyons can be over 3 kilometers deep. Sprawling **abyssal fans** (deep-sea fans) are often found at the mouths of submarine canyons (Figure 51). Abyssal fans resemble alluvial fans in shape. The canyons are sometimes eroded by the constant movement of sand by longshore currents, which also erodes the underlying bedrock. Sand movement that is quite rapid and steep is termed a **sand fall.** Regular bottom currents streamline and shape abyssal fans.

Submarine Canyons

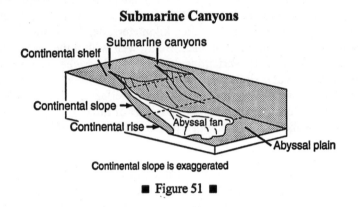

■ Figure 51 ■

Turbidity currents. Turbidity currents are large volumes of dense, sediment-laden water that result when sand and mud on the continental slope are dislodged by landslides or earthquakes and become suspended in the water. The turbidity currents are denser than water and behave as a separate flow that rolls down the slope at speeds up to 60 kilometers per hour. The resulting sedimentary deposits are called **turbidites.** Although large-scale turbidity currents have never been witnessed, they are likely the culprits that periodically break telephone cables on the ocean floor. Both graded beds and shallow-water fossils found in abyssal plain sediments are also indirect evidence that large-scale turbidity currents exist.

Ocean Floor Sediments

There are three kinds of sea floor sediment: terrigenous, pelagic, and hydrogenous. **Terrigenous sediment** is derived from land and usually deposited on the continental shelf, continental rise, and abyssal plain. It is further contoured by strong currents along the continental rise. **Pelagic sediment** is composed of clay particles and microskeletons of marine organisms that settle slowly to the ocean floor. Some of these organic sediments are called calcareous or siliceous "oozes" because they are so thick and gooey. The clay component (or sometimes volcanic ash) is generally carried from land by wind and falls on the surface of the ocean. Pelagic sediment is least abundant on the crest of midoceanic ridges because of the active volcanism. **Hydrogenous sediments** are rich with minerals, such as manganese nodules, that precipitate from seawater on the ocean floor.

Active Continental Margins

Active continental margins are those that are tectonically active, such as along much of the Pacific coast. Active margins are marked by earthquakes, volcanoes, and mountain belts. Unlike passive margins, they lack a continental rise and abyssal plain. Instead, the continental slope ends in an oceanic trench, and beyond the trench, the topography is hilly and irregular, often dotted with rugged volcanic seamounts.

An **oceanic trench** is a narrow trough parallel to the coastline that can reach a depth of 10 kilometers (6 miles) or more and a slope of 15 degrees. Volcanoes and earthquake-producing Benioff zones are associated with oceanic trenches (Figure 50). These trenches, such as the Mariana Trench, are the deepest features of the oceans.

Passive Continental Margins

Passive continental margins develop along coastlines that are not tectonically active, including much of the Atlantic Ocean coastline. Many passive continental margins have a **continental rise,** a very low-angle ridge of sediment that forms between the continental slope and the abyssal plain (Figure 49). The sediments that form the continental rise are deposited on oceanic crust by turbidity currents, **contour currents,** which flow parallel to the edge of the continental slope, and regular ocean currents.

Abyssal plains are flat-lying expanses of horizontally deposited sediment that accumulates on the ocean floor at the base of the continental rise. The abyssal plain can be very thick and completely bury large-scale topographical features of the ocean floor. Some sections show graded bedding that indicates deposition from turbidity currents. Abyssal plains are some of the flattest features in the world.

Reefs

Reefs are accumulations of organisms that form in warm, shallow ocean environments. Typically consisting of coral and algae, reefs are resistant ridges that rim islands, lagoons, and other shorelines (Figure 52). **Fringing reefs** are flat expanses of reef that grow in the shallow water near the shore. **Barrier reefs** are elongate features that develop offshore parallel to the coastline and are separated from the coastline by deep lagoons. **Atolls,** circular reefs found in deeper water, are the result of reef development around the flank of a volcano that has since subsided but to which the corals are still anchored.

Reefs

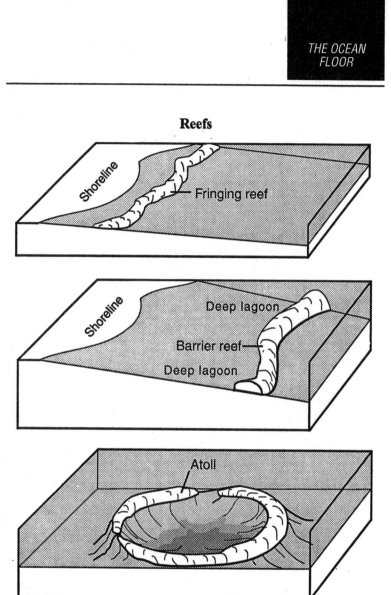

■ Figure 52 ■

Midoceanic Ridges

Oceanic ridges are some of the longest and steepest features in the world. They form an almost continuous mountain chain that is about 75,000 kilometers (45,000 miles) long and 3 kilometers (nearly 2 miles) high. The ridges form along deep crustal faults that separate the ocean floors into approximately two equal halves. The outlines of the midoceanic ridges generally parallel those of the continental coastlines on either side of the ridge (Figure 53).

A Midoceanic Ridge and Parallel Coastlines

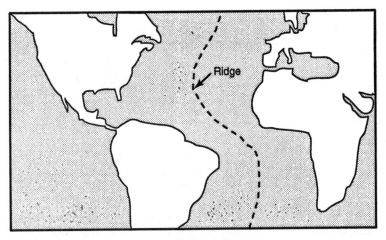

■ Figure 53 ■

A tensional graben (rift valley) runs down nearly all the midoceanic ridge crests. Shallow earthquakes and abnormally high heat flows are associated with ridges. New oceanic crust is produced in the divergent plate-boundary environment of the ridge. Mostly basaltic, these rocks are pushed away to the sides as more magma erupts and cools at the crest (Figure 54).

Crust Formation at a Midoceanic Ridge

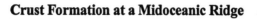

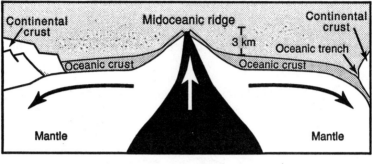

■ Figure 54 ■

The midoceanic ridges are cut by approximately perpendicular fracture zones that offset segments of the ridge (Figure 55).

Midoceanic Ridge Fracture Zones

■ Figure 55 ■

A **transform fault** is that portion of a fracture between two offset portions of the ridge (Figure 56). Transform faults also host shallow-focus earthquakes. The faults are equilibrium features that result from a curved midoceanic rift that is "trying" to diverge evenly on both sides. At transform faults, the segments of crust are moving in opposite directions.

Midoceanic Ridge Transform Faults

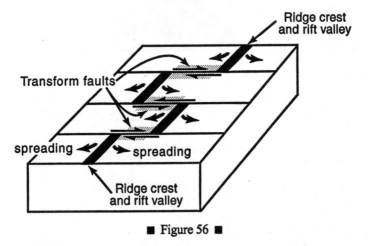

■ Figure 56 ■

Seamounts are conical volcanic peaks on the ocean floor. Usually basaltic in composition, they are at least 1,000 meters high and typically form near midoceanic ridges. They can sometimes rise above sea level to form islands. **Guyots** are submerged, flat-topped seamounts that were once at sea level and eroded by continual wave action. Chains of seamounts and guymots are called **aseismic ridges** because they are not associated with earthquakes.

Oceanic Crust

Seismic data, dredging, and drilling have shown that the oceanic crust is divided into three parts. The top layer consists of terrigenous or pelagic sediment that averages about 0.5 kilometer in thickness. These are typical sediments from the abyssal plain. The middle layer is composed of pillowed basalt flows and underlying, steeply dipping basaltic dikes. It is thought the dikes were the feeders that supplied the lava for the overlying flows. The middle layer is about 2 kilometers thick. The bottom layer is about 5 kilometers thick and thought to consist of swarms of basalt dikes and sill-like gabbro intrusions. This bottommost section of oceanic crust lies on top of the mantle.

This interpretation for the bottom layer is based on the detailed study of **ophiolites,** which are mafic rock sequences at the earth's surface that are believed to be pieces of ancient oceanic crust that were thrust *onto* the continent (**obduction**) during mountain-building. Larger, more complete ophiolite sequences show the same general layering as that shown by drilling and seismic studies of the ocean crust: pillowed lavas, vertical basalt dikes, and massive, coarse-grained gabbro. The combined thickness of the dikes and the gabbro units in big ophiolite complexes is about the same thickness as that indicated by seismic studies for the bottom layer.

In a complete ophiolite sequence the gabbro may be underlain by ultramafic rocks such as peridotite. If ophiolites are truly oceanic crust, this contact could then represent the Moho. (See p.131.) The peridotite would then be a segment of the uppermost part of the mantle.

Magma generation, igneous intrusions, metamorphism, volcanic action, earthquakes, faulting, and folding are usually the result of **plate tectonic** activity. The earth's crust is divided into six large pieces, and about twenty smaller pieces, by deep fault systems. These **crustal plates** include both oceanic and continental crust. Underlying convection currents in the mantle and lower crust are thought to create forces that push and pull these plates at the surface. Intense geologic activity occurs where plates move apart (divergent boundaries), collide (convergent boundaries) or slide past one another (transform boundaries). About 200 million years ago, it is thought, plate tectonic forces began to break a single continental land mass into pieces that spread apart to form the continents as we know them today.

Early Evidence for Plate Tectonics

Continental drift. As world maps began to improve in the 1600s and 1700s, scientists noticed that the continents, especially South America and Africa, would roughly fit together like the pieces of a jigsaw puzzle were they in contact with one another (Figure 57). The idea that the continents were once joined together and somehow split apart was originally called **continental drift** and was the precursor to modern-day plate tectonic theory. As more was learned over the centuries (especially the existence of deep midoceanic rifts that parallel the outlines of the continents), the idea of plate tectonics became more and more plausible to geologists.

Formation of the Continents

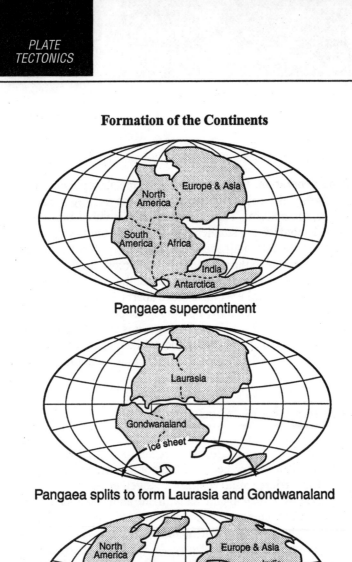

Pangaea supercontinent

Pangaea splits to form Laurasia and Gondwanaland

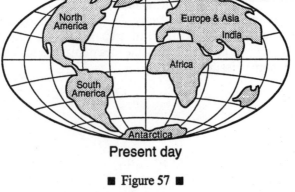

Present day

■ Figure 57 ■

The work of Alfred Wegener. Scientists began to talk seriously about continental drift in the mid 1800s. **Alfred Wegener,** a German climatologist, noticed that certain mountain belts, rock formations, strikes and dips, and fossil remains were nearly identical in parts of South America, Australia, India, and Africa. He reasoned that if a shared species such as *Mesosaurus* survived swimming the oceans between the continents, their remains should be widely distributed in oceanic sediments—yet they have been found only in eastern South America and southern Africa. Considering the distance between the continents, Wegener concluded that to have the same unique fossil assemblages they had to have been part of the same larger landmass. He named this theoretical supercontinent Pangaea, which, he suggested, split to form Laurasia and Gondwanaland. Laurasia, the northern part, later fragmented again to form North America and Eurasia. Gondwanaland broke apart to form South America, Africa, India, Australia, and Antarctica.

Wegener's studies also revealed that a well-defined period of late Paleozoic glaciation affected the southern Gondwanaland continents. If the continents had been at their present positions and covered by the same ice sheet, the weather would have been cold enough to result in glaciation of the northern continents; however, late Paleozoic climates in North America and Europe were actually warm and humid. The occurrence of glacial striations (and the directions of ice movement) on the southern continents strongly suggest Gondwanaland was a single landmass toward the end of the Paleozoic era. The ice sheet was centered over present-day Antarctica and spread westward over part of South America, north and westward into Africa, and eastward into India and Australia, forming a radial pattern.

Further intrigued, Wegener studied rocks around the world to reconstruct the climate zones for each geologic time period. For example, limestones and reefs indicate warm ocean waters near the equator, and glacial deposits would indicate colder climates. Wegener discovered that the positions of the North and South Poles in the geologic past were quite different from their positions today, at least in their relationship to the positions of the continents. For example, fossil trees from coal fields in frozen terrain like Siberia contain no growth rings, indicating they grew very rapidly in a tropical climate.

This evidence for **polar wandering** meant that either the geographical poles moved and the continents were stationary or the continents moved and the geographical poles remained stationary.

Paleomagnetic Evidence for Plate Tectonics

In the 1940s and 1950s, technology had advanced to the stage at which **paleomagnetic fields** from the geologic past could be measured with some reliability from rocks. Just as Wegener's geologic work identified where the geographical poles had been in the geologic past, that of geophysicists was starting to determine where the magnetic poles had been located. The alignment of a magnetic mineral in a cooled igneous rock points to the magnetic north pole, and the dip of the mineral reveals how far the rock formed from the pole. The paleomagnetic evidence revealed that the magnetic poles also had different locations relative to the continents than they do today. Magnetic minerals on one continent do not point to the same pole position as do those from the same time period on another continent. This would suggest either that there were multiple north poles during the same time period or that the continents moved in relation to a single north pole. Geophysicists concluded that the magnetic poles remained stationary, and the continents, after splitting apart, diverged along different paths.

Sea Floor Evidence for Plate Tectonics

The technologies developed in the 1940s and 1950s also permitted more detailed mapping of the ocean floor and continental margins. A much better fit between the rifted continents is apparent when the shape of the continental slope is used instead of the continent's shoreline. Detailed mapping of distinctive rock units that extend out to sea along the South American and African coasts and North American and British coasts has shown that they would converge perfectly if the continents could be fitted together.

Sea floor spreading. In the 1960s, geologist **Harry Hess** proposed that the sea floor was moving outward from the midoceanic ridges. His theory of **sea floor spreading** maintained that new basaltic oceanic crust forms at a midoceanic ridge and is slowly pushed away on both sides toward the continents as more new crust is produced. (Measurements indicate that new crust moves away from a ridge at rates from 2 to 10 cm/year.)

A midoceanic ridge is called a **spreading axis** or **spreading center.** **Subduction** is the process by which the oceanic crust is pushed against, and finally underneath, continental or oceanic crust. Subduction zones are often marked by overlying chains of volcanic islands called **island arcs.** Geologists believe sea floor spreading results from convection in the mantle and lower crust that brings hotter, less dense, and more plastic material up toward the surface; the colder, more dense rock and sediment, such as subducted crustal material, sinks toward the mantle (Figure 58). These convective forces tear the ocean crust apart at the midoceanic ridge, forming a rift valley marked by high-angle faults, basaltic lavas, and high heat flows. The Mid-Atlantic Ridge is one of the best studied midoceanic ridges. It separates North America from Europe. Its 10,000-foot-tall mountain peaks lie about a mile below the surface of the ocean.

Sea Floor Spreading

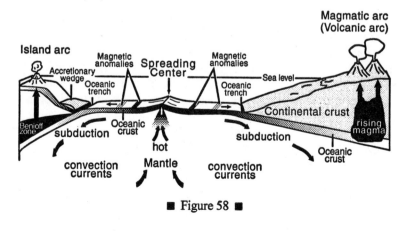

■ Figure 58 ■

As the hot mantle rock ascends toward a midoceanic ridge, it cools and starts to move laterally away from the ridge. This mantle movement drags the overlying oceanic crust along with it. The mantle material continues to cool, and eventually begins to sink. At this point, the oceanic crust begins to be subducted in deep oceanic trenches. Ocean trenches develop over the subduction zone where the subducted plate begins to bend and descend. Essentially great troughs, these are some of the deepest features on Earth, averaging about 10 kilometers in depth. The progressive cooling trend away from the ridge is illustrated by the low heat flows that have been measured at oceanic trenches. Subduction causes contact and friction with the overlying plate, which results in andesitic volcanoes and earthquakes along dipping Benioff zones.

The youngest oceanic crust is formed at the crest of a midoceanic ridge, and the crust becomes progressively older away from the ridge. The oldest oceanic crust is then subducted and reassimilated in the mantle. These events explain why oceanic crust is not as old as many continental rocks. Once cooled, the new crust is broken in half along the rift valley, each half moving away from the rift as new crust is formed. The oldest ocean crust is only about 150 to 200 million years old.

Magnetic anomalies. Magnetic surveys over the ocean floor in the 1960s revealed symmetrical patterns of **magnetic "bands,"** anomalies parallel to midoceanic rifts (Figure 58). The same patterns in relation to midoceanic rifts are present in different oceans. The magnetic anomalies coincide with the episodes of magnetic reversals that have been documented from studies on land, indicating that the andesitic rocks that form new oceanic crust in the tensional setting of the rift valley record the earth's magnetic field as they cool. A rock has a **normal (positive) polarity** when its paleomagnetic field is the same as the earth's field today. The positive magnetism adds to the earth's magnetic field and creates a higher magnetic measurement at that location. Rocks are **negatively polarized** when the earth's field is reversed, which reduces the earth's net field strength. Since the ages of these anomalies are known from dating the paleomagnetic reversals on land, the rate of movement of the ocean floor can be calculated. The fact that new ocean crust moves away from the midoceanic ridge at speeds that range from 2 to 10 centimeters per year has also been documented using satellite measurements and radar. For example, if it is known that a segment of sea floor that formed 10.0 million years ago is now 50 kilometers (5.0 million cm) away from the crest of the ridge, it can be calculated that it traveled that distance at about 2 centimeters per year. By using the calculated ages for episodes of paleomagnetic reversal, scientists can construct sea floor age maps, which confirm that the youngest oceanic crust is presently being formed at midoceanic ridges and that the oldest is about 150 to 200 million years old, or late Jurassic in age. This older material is the farthest from the spreading centers and is the next crust to be subducted. Sea floor age maps have been proven correct by the age dates calculated from hundreds of rock samples gathered from the ocean floor.

Seismic studies. More proof for sea floor spreading comes from seismic studies indicating that earthquakes occur along the rift valley of a midoceanic ridge and the cross-cutting fractures that offset it. Rift valley earthquakes occur only along transform faults, those portions of the fracture zone located between the offset sections of a ridge and rift valley. Because of the way in which the sea floor spreads (that is, away from both sides of a midoceanic ridge), transform faults are the only areas along the fracture zone in which sections of the oceanic crust pass one another in opposite directions (Figure 56). The concentration of earthquakes in the transform-fault sections of the fracture zones further supports the concept of ocean crust moving away from a midoceanic ridge.

Modern plate tectonic theory. By the 1960s, the theories of continental drift and sea floor spreading were supported by reliable scientific data and combined to develop modern-day **plate tectonic theory.** The theory maintains that the crust and uppermost mantle, or lithosphere, is segmented into a number of solid, rigid slabs called

---

PHYSICAL GEOLOGY

lithospheric plates. These slabs move slowly over the asthenosphere, the 200-kilometer-thick zone of more plastic mantle material that underlies the plates. New oceanic crust is created at the crests of the midoceanic ridges and pushed laterally away by new accumulations of crust. It begins to cool as it moves away from the high heat flows at the ridge. By the time it is subducted at the convergent boundary with another plate, it is cold and dense enough that it begins to sink back into the mantle. Subduction is also probably a function of a down-turning mantle convection current below the converging plates.

Detailed geologic mapping on the continents has shown that crustal plates have been in motion, collided together to form new continental masses, and broken apart again many times in the past two billion years, and probably longer. Reconstructions of the paleoterrains in the geologic past reveal the same spatial relationships between terrain types that we see today along convergent, divergent, and transform boundaries. Suture zones that represent the lines along which separate crustal masses collided and joined together can be mapped in the field. Pangaea, the supercontinent that existed about 200 million years ago, consisted of other continental masses that had been welded together during previous plate tectonic collisions.

In summary, the compelling evidence for plate tectonic theory is

- the occurrence of most earthquakes, volcanic activity, and mountain-building along plate boundaries

- the fit of continental edges

- the match of specific geologic features between continents, such as rock formations and fossils

- glacial striations on the southern continents

- the apparent variety of geographic pole locations

- the variety of magnetic pole locations

- sea floor spreading

- the fact that midoceanic ridges parallel continental coastlines

- the symmetry of banded magnetic anomalies on the sea floor

- earthquake activity along transform faults

- "young" age dates of ocean floor rocks

- satellite measurements that indicate changes in distance between known points on different plates.

How Plates Move

Most plates consist of both continental and oceanic crust. Plates move away from each other at spreading centers (divergent boundaries). A convergent plate boundary separates plates that are moving toward each other. A transform plate boundary is a fault zone along which two plates slide in opposite directions. Oceanic crust is subducted underneath continents or in oceanic trenches. Continental crust is less dense than oceanic crust and therefore will not subduct because it is lighter.

Compared to its edges, a plate's interior is relatively stable, with few earthquakes and little igneous activity or structural deformation. Flood basalts and mantle plume "hot spots" have been known to occur in the interior, but the majority of seismic, volcanic, and mountain-building activity occurs along a plate's boundaries (Figure 59), including frequent earthquakes.

Several Tectonic Plates

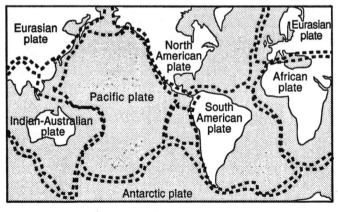

- Figure 59 -

Types of Plate Boundaries

Convergent boundaries. Plates may converge directly or at an angle. Three types of **convergent boundaries** are recognized: continent-continent, ocean-continent, and ocean-ocean.

Continent-continent convergence results when two continents collide. The continents were separated at one time by oceanic crust that was progressively subducted under one of the continents. The continent overlying the subduction zone will develop a magmatic arc until the ocean floor becomes so narrow that the continents collide. Because the continents are less dense than the oceanic crust, they will not be pulled down the subduction zone. One continent may override the other for a short distance, but the two continents eventually become welded together along a geologically complex suture zone that represents the original line of collision. The crust is thickened along the suture zone, resulting in isostatic uplift, mountain-building, and thrust faulting.

Ocean-continent convergence occurs when oceanic crust is subducted under continental crust. This forms an **active continental margin** between the subduction zone and the edge of the continent. The leading edge of the continental plate is usually studded with steep andesitic mountain ranges. Earthquakes occur in the Benioff zones that dip underneath the continental edge.

Magmatic arc is a general term for belts of andesitic island arcs and inland andesitic mountain ranges that develop along continental edges. These mountain ranges (also called volcanic arcs) are underlain by crust that has been thickened by intrusive batholiths that were generated by partial melting along the underlying subduction zone. The Sierra Nevada in California and Nevada is a volcanic arc. Volcanic arcs result from isostatic processes, compressional forces along the leading edge of the continent, and thrust faults that move slices of mountain-belt rocks inward over the continental interior, creating **backarc thrust belts.** The additional weight of these rocks downwarps the inland area, forming a **foreland basin.** The foreland basin fills with eroded material from the mountain ranges or occasionally with marine sediments if it becomes submerged.

Ocean-ocean convergence occurs when two plates carrying ocean crust meet. One edge of ocean crust is subducted beneath the other at an ocean trench. The ocean trench curves outward toward the subducting plate over the subduction zone. Data from earthquakes along the subducting plate show that the angle of subduction increases with depth. Subduction probably occurs to a depth of at least 670 kilometers (400 miles), at which point the plate probably becomes plastic.

Andesitic volcanism often forms a curved chain of islands, or island arc, that develops between the oceanic trench and the continental landmass. Modern-day examples of island arcs are the Philippines and the Alaska Peninsula. Geologists think that at a depth of about 100 kilometers (60 miles) the asthenosphere just above the subduction zone partially melts. This mafic magma may then assimilate silicious rocks as it moves up through the overlying plate, forming a final andesitic composition that vents to form the island arc. The distance the island arc forms from the oceanic trench is dependent on the steepness of the subduction zone—the steeper the angle of subduction, the more quickly the subducted material reaches the magma-forming depth of 100 kilometers, and the closer the arc will be to the oceanic trench.

The trench becomes filled with folded marine sediments that slide off the descending plate and pile up against the wall of the trench. This accumulation is called the accretionary wedge or subduction complex. The accretionary wedge is continuously pushed up to form a ridge along the surface of the trench over the subducted crust. The forearc basin is the relatively undisturbed expanse of ocean floor between the accretionary wedge and island arc; the area on the continental side of the arc is called the backarc.

The backarc basin, the basin that occurs between the island arc and the continental mass, is occasionally split by new extensional forces into two parts that migrate in different directions (backarc rifting). In other words, a "mini" spreading center develops as an equilibrium response to changes in the way the plate is being subducted. This backarc spreading can push the island arc away from the continent toward the subduction zone. If it develops along the continental edge, it can also split off a strip of the continent and push it seaward toward the subduction zone—Japan is a modern-day example. The rifting may be caused by a mantle plume that has come near to the surface and is spreading out, creating convection currents that stretch the crust to the point of breakage.

The locations of oceanic trenches shift gradually with time, a phenomenon thought to be caused by the force of the leading edge of the overlying plate, which pushes the trench back over the subducting plate. This is because the overlying plate has a forward tectonic force and a gravitational force that bears down on the subducting plate. Some geologists believe the subducting material sinks at an angle that is steeper than that of the subduction zone, which would tend to pull the subducting plate away from the overlying plate, allowing the overlying plate to again move forward and push the oceanic trench back over the subducting plate.

Divergent boundaries. A **divergent plate boundary** is formed where tensional tectonic forces result in the crustal rocks being stretched and finally split apart, or rifted. The central block drops to form a graben, and basaltic volcanism is abundant along the rift's faults. The rise of hot mantle material beneath the rift zone pushes the rift valley farther apart (Figure 60). Today's active divergent boundaries are midoceanic ridges (sea floor spreading centers). Divergent boundaries can also develop on land, as did those that broke up Pangaea about 200 million years ago. Continental rifting can end before the crustal mass has been fully separated. These **failed rifts** then become seas or large basins that fill with sedimentary material. An example of a failed rift is the approximately two-billion-year-old midcontinental rift in the United States, which extends from the Great Lakes area southward to below the Great Plains. The rugged topography of the rift was filled with coarse-grained sediments and volcanic flows and has since been buried by thousands of feet of sedimentary rock deposited under Paleozoic oceans.

Geologists have debated for years whether uplift causes rifting or whether rifting causes uplift. Some scientists feel that rifting thins the crust, reducing the amount of pressure it can exert; the reduced pressure allows deeper, more pressurized rocks to ascend, causing uplift (similar to unloading and dome structures). Most geologists agree that uplift occurred after the rifting that resulted in the Red Sea in the Middle East.

Divergent Plate Development

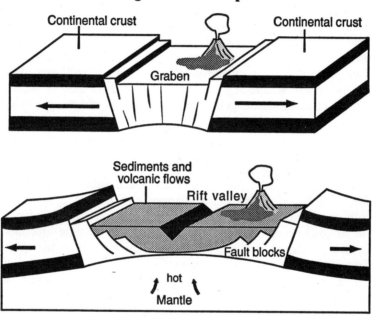

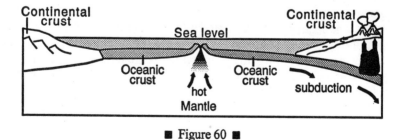

■ Figure 60 ■

Eventually the crust is totally split by continued divergence along the rift, and the two parts are separated by a new sea that floods the rift valley. New, basaltic oceanic crust continues to build up along the rift, causing high heat flows and shallow earthquakes. The Red Sea is at this stage of divergent separation.

Rivers do not discharge into the new ocean because the continental edges have been uplifted by the rising mantle material and slope away from the ocean. As divergence continues, the sea widens and the midoceanic ridge continues to grow. Eventually the continental edges subside as the underlying rocks cool and are further lowered by erosion. Rivers begin to flow into the sea forming deltas, and marine sedimentation begins to form the continental margin, shelf, and rise.

Transform boundaries. A **transform boundary** is a fault or a series of parallel faults (fault zone) along which plates slide past each other via strike-slip movements. As previously discussed, transform faults connect offset midoceanic ridges (including the rift valleys). The motion between the two ridge segments is in opposite directions; beyond the transform fault, crustal movement is strike-slip in the same direction. Thus, the transform fault "transforms" into a fault that has different motions along the same fault plane. Transform faults can connect diverging and converging boundaries or two converging boundaries (such as two oceanic trenches). Transform faults are thought to form because the original line of divergence is slightly curved. As an adjustment to mechanical constraints, the tectonic forces break the curved or irregular plate boundary into a series of pieces. The segments are separated by transform faults that are parallel to the spreading direction, allowing the ridge crest to be perpendicular to the spreading direction, which is the easiest way for two plates to diverge. Transform faults allow the divergent boundary to be in a structural equilibrium.

Why Plates Move

Both plate boundaries and plates move over time. As previously described, plates can change the locations of trenches and subduction zones, as well as the positions of midoceanic ridges and transform faults. For example, subduction at a convergent boundary can stop in one location and begin nearby in another. Plates can become larger or smaller over time depending on the generation rates of new crust at spreading centers and the rates of subduction.

Convection currents. Some geologists favor convection currents in the mantle as the best explanation for plate tectonic movement. It is reasonable to assume that the heat radiated from the core creates convection currents in the mantle, and the mantle rocks begin to move plastically. Convection movement in the uppermost layers of the mantle may pull on the lithospheric rocks, breaking them into huge plates that move slowly on the more plastic, lubricated surface of the asthenosphere. Another possibility is that the 670-kilometer boundary in the mantle breaks the convection pathways into an upper and lower part; convection in the lower part may induce the convection currents in the upper part (less than 670 kilometers deep) that move the plates. Still others believe that plate motion on the surface *creates* the underlying mantle convection—that is, when plates diverge, hotter mantle rock rises upward to fill the space between the plates, helping to push them apart; as the plates move away from the spreading center they cool and begin to sink, creating downward currents.

Mechanisms of plate movement and subduction. Three mechanisms (in order of importance) have been proposed to explain why plates move apart and subduct: slab-pull, ridge-push, and trench-suction. **Slab-pull** is the result of a plate subducting at a steep angle through the mantle; this downward motion tends to pull the other side of the plate away from the ridge crest. The **ridge-push theory** maintains that new crust cools as it moves away from the ridge, becoming more dense, sinking, and forming a slope on the midoceanic ridge. A parallel slope may develop underneath the plate at the base of the lithosphere. These surfaces are zones of weakness that help the plate move away from the ridge. When a plate is subducted at a steep angle, it also creates **trench-suction** that pulls the overlying plate, and the trench, toward the ridge.

Mantle Plumes

Hot mantle rock that rises toward the earth's surface in a narrow column is called a **mantle plume.** Plumes can be located beneath continental or oceanic crust or along plate boundaries. Plumes are thought to spread out laterally at the base of a continent, creating increased pressure that stretches the crust and results in uplift, fracturing, rifting, or flood basalts. Mantle plumes are thought to be strong enough to induce rifting and the formation of plates. The pressure creates a domed region that eventually splits in a three-pronged pattern (**triple junction** or **triple point**). If rifting continues, two of the three faults become active, forming the continental margins of two new continents. The two faults join to form an active divergent boundary that dissipates the tectonic forces. The third "arm" becomes a **failed rift,** or **aulacogen,** that rapidly fills with sediment.

The best example of a triple junction in the world is provided by the three faults marked by the Red Sea, the Gulf of Aden, and the inactive African Rift Valley. The rifting is separating the Arabian Peninsula from the African continent and is thought to be related to a mantle plume. Other areas that are underlain by mantle plumes are the Hawaiian Islands (oceanic crust) and Yellowstone National Park (continental crust) in the United States.

Pangaea

Pangaea is the supercontinental landmass recognized by geologists to have been separated by plate tectonic activity. Previous plate tectonic collisions welded other continental masses together to form Pangaea.

Evidence including paleomagnetism and the correlation of unique geologic features indicates that Pangaea began to break apart about 200 million years ago (Figure 57). Two east-west trending rift zones separated Pangaea into two parts: to the north, **Laurasia** (North America and Eurasia) and to the south, **Gondwanaland** (modern-day South America, Africa, India, Australia, and Antarctica). The rifting event was marked by a huge outpouring of basaltic lava flows. After 20 million years of rifting, these two land masses were separated by narrow oceans formed by the inpouring of the Tethys Sea (including the newborn Atlantic Ocean). During the Jurassic period, about 135 million years ago, more rifting began to fragment the two land masses into the continents we know today. By 65 million years ago, the mid-Atlantic oceanic ridge and associated transform faulting was well developed, South America had fully separated from Africa, and North America was beginning to drift from Europe.

Mountains result from the application of tectonic forces to rocks, usually sedimentary or volcanic rocks. (These may be changed to metamorphic rocks as mountain-building progresses, and at times metamorphic rocks can be pushed into mountains). **Mountain-building** on continents is associated with intense deformation, folding, and faulting, usually along convergent plate boundaries (Figure 61). An **orogeny,** or **orogenesis,** is the overall process by which a mountain system is built.

Mountain-Building

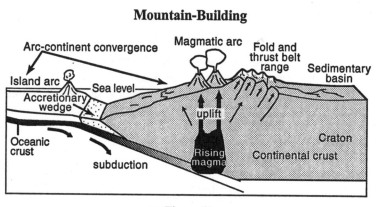

■ Figure 61 ■

Mountain ranges are groups of mountain peaks or ridges that form discrete topographic areas that are usually bordered by valleys or rivers. It takes tens or hundreds of millions of years to develop **mountain belts,** long chains of mountain ranges that can extend across continents or along their edges.

Much of our understanding of mountain-building processes has come from studying ancient mountain belts that have since eroded to form flat erosional surfaces. Fieldwork in mountain ranges can be difficult because the rocks are complexly folded and faulted and because of the great changes in elevation. But such fieldwork can be productive because these features are important in the mapping and examination of the third dimension of the earth.

Features of Mountain Belts

Mountain belts typically are thousands of kilometers long and hundreds of kilometers across and parallel continental coastlines. The American Cordillera is a series of steep mountain ranges that rim the western edge of North and South America; it is one of the longest mountain belts in the world. In general, the taller mountains are geologically younger than lower mountains (for example, the steeper Rocky Mountains are younger than the lower and more rounded Appalachian Mountains) because older ranges have undergone more weathering and erosion. Most mountain ranges are uplifted, erode to low elevations, and are uplifted again before they become stable.

Major mountain ranges in the United States include the Appalachian Mountains, the Rocky Mountains, the Ozark Mountains, and the many ranges along the West Coast. Fossil evidence and age dating indicate the rounded hills of the Appalachian and Ozark Mountains are some of the oldest mountains in the United States.

Cratons. Billions of years ago the now-stable interior of North America was a mountainous, tectonically active region that eventually stabilized and weathered to a **peneplain** (an area reduced by erosion nearly to a plain). A continental interior that has been structurally inactive for hundreds of millions of years is called a **craton.** It is composed of mostly plutonic and metamorphic rocks. The craton is a "basement" upon which sequences of sedimentary rocks were deposited under marine or nonmarine conditions. The central United States is covered by about 2,000 meters of sedimentary rocks that were deposited in shallow Paleozoic oceans. Continents have grown larger through **accretionary episodes** in which mostly sedimentary material and volcanic arcs were welded to the craton through plate collisions, usually resulting in mountain-building.

Rock types. Mountains are typically composed of folded sedimentary strata that may be up to five times as thick as the original sedimentary sequence that covered the cratonic interior. The folded and broken layers indicate the rock has undergone deformation during mountain-building. Since mountain belts typically form along tectonically active coastlines and above subduction zones, much of the sedimentary rock is marine in origin. The sediments are often parts of the accretionary wedge that have been compressed, folded, and driven onto the continent by plate tectonic processes.

How intensely a mountain belt is folded depends on how great the tectonic forces were. Mountain-building forces are intensely compressional, and the sedimentary sequence in a basin is often squeezed into a mountain range that is less than half the width of the original basin. Rock layers are typically contorted into tight fold patterns, including overturned or recumbent folds. **Fold and thrust belts** in many mountain ranges are the result of multiple thrust layers (sheets) of rock that have been thrust forward and stacked vertically along the low-angle **detachment faults** that separate the thrust sheets. After uplift has been completed, a later stage of tensional stress develops that forms a series of fault-block (horst and graben) mountains. The faulting is an adjustment to the extensional stress created by the vertical uplift.

The core of a mountain range tends to be its most intensely metamorphosed part. The metamorphic rocks were originally sedimentary rocks or volcanic rocks that were intensely metamorphosed through deep burial, folding, and tectonic uplift. It is often difficult to recognize the original rock types, and metamorphic rocks are typically mapped as "schist" or "gneiss." **Migmatites** are some of the most intensely metamorphosed rocks that are found in the cores of mountain ranges. The large batholithic intrusions that underlie mountain ranges were formed by partial melting during the mountain-building process. The continental crust under mountain ranges is thicker than that under the cratonic interior; similarly, the crust under younger mountain ranges is thicker than the crust under older ranges. The uplift of these blocks of crust eventually stabilizes through isostatic adjustments. Geologically young, tectonically active mountains have more earthquakes and volcanic activity than the older, more stabilized mountain chains.

Types of Mountains

Although each mountain range has a unique set of characteristics, a particular mountain can be structurally classified as upwarped, volcanic, fault-block (horst and graben), or folded (complex) (Figure 62). It is not unusual to find all four categories of mountains within a single mountain range.

Types of Mountains

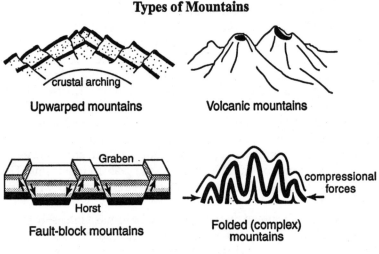

Upwarped mountains   Volcanic mountains

Fault-block mountains   Folded (complex)
                              mountains

■ Figure 62 ■

Upwarped mountains are generally the result of broad arching of the crust or sometimes great vertical displacement along a high-angle fault. The Black Hills in South Dakota and the Adirondack Mountains in New York are upwarped mountain ranges. These mountains are more rounded and show some unloading features such as exfoliation. **Volcanic mountains** are the accumulations of large amounts of volcanic lavas and pyroclastic material around the volcanic vent, such as seamounts and stratovolcanoes. The Hawaiian and Aleutian Islands are volcanic mountains. **Fault-block mountains** result from tensional stress. They are bounded by high-angle normal faults, and usually form a series of horsts and grabens. Broad crustal uplift (possibly a result of subduction stresses or mantle upwelling) can stretch and break the crust, creating fault zones along which the blocks move or slide. Uneven tectonic uplift can tilt the blocks. A good example of fault-block mountains are those in Nevada that are part of the Basin and Range region. **Folded,** or **complex, mountains** are created by intense compressional forces that fold, fault, and metamorphose the rocks, resulting in many of the world's biggest mountain belts, such as the Himalayas.

How Mountains Form

In general, it takes hundreds of millions of years for mountain belts to form, stabilize, and erode to become part of a stable craton. This evolution is marked by three stages: accumulation, orogeny, and uplift/block-faulting.

Accumulation. Many mountains contain sequences of sedimentary and volcanic rocks that reach thicknesses of 2,000 to 3,000 meters. Most of this material was deposited in a passive or active continental marine environment during the **accumulation stage.** The sedimentary material typically weathers from the continental landmass or offshore island arc; deep marine sediments can also be scraped from the subducting plate and piled onto the accretionary wedge. Thick sequences of sandstone, shale, and limestone with minor volcanic material accumulate along passive continental margins, such as the eastern coast of the United States. Sediments that accumulate along a convergent boundary (active continental margin) are more varied than those along a passive margin and often contain up to 50 percent andesitic flows and tuffs. Limestones are rare to absent. Graywackes are common and represent a rapid accumulation of sediment from a nearby magmatic arc. These sedimentary and volcanic deposits along the continental margins have been pushed up into many of the mountain ranges we see today.

Mountain-building convergence. Orogenesis is the mountain-building and associated folding, faulting, deformation, and metamorphism that result from the onset of intense tectonic stress. Igneous intrusions are also common. The layered rocks are tightly compressed into folds that often result in thrust faulting. The deepest rocks are metamorphosed into schists, gneisses, and migmatites. Compression also results in vertical uplift of the deformed rock sequence. Block faulting can also occur after the forces have thrust the metamorphosed and deformed rocks upward and outward.

Ocean-continent convergence deforms the accretionary wedge, metamorphoses rocks in the subduction zone, creates a mountainous magmatic arc, and develops fold-and-thrust belts on the backarc side of the magmatic arc. Ocean-continent convergence results in the formation of the rugged topography of the deep ocean trenches and seamounts. The magmatic arc is elevated because of the massive igneous upwellings underneath it.

Arc-continent convergence results when the intervening ocean is destroyed by subduction, welding an island arc to the continental edge. This convergence also results in deformation, uplift, and orogeny (Figure 61). Tectonic forces along the continental edge continue to generate heat, igneous intrusions, and compressional forces to the continental edge. It is probable that the northwestern United States was formed by a series of arcs that collided with and were welded to the North American craton.

The collision of two continental masses, or **continent-continent convergence,** also results in the formation of mountain belts. The thick sedimentary sequences that formed on both continental edges are squeezed into intensely deformed mountain ranges that are some of the highest in the world, such as the Himalayas between India and the rest of Asia. The Himalayas are still rising because of continued compression along the suture boundary, and they host frequent earthquakes. Some geologists theorize that the Appalachian Mountains in the eastern United States were built as a result of the collision of the European and African plates with the North American plate that helped form the supercontinent Pangaea. The Appalachians and the Caledonian Mountains of Great Britain and Norway were all once joined along the same suture zone before Pangaea rifted into the continents that we see today.

Postconvergence mountain-building. A mountain range undergoes additional **uplift** and **block-faulting** after orogeny has ceased. Because the continental crust was thickened during mountain-building, the gradual uplift over tens of millions of years is a result of isostatic adjustment. As material is eroded from the mountain belt, more uplift compensates for the loss of weight (mass) during erosion. The uplift creates vertical and extensional (tensional) stresses that result in the block-faulting of mountain ranges along a series of normal faults. Fault blocks may also be tilted if the stresses are unevenly distributed. Scattered volcanic activity can also be part of this phase of mountain development. Fault-blocked mountain ranges are usually separated by valleys filled with thousands of feet of sediment, such as those of the Great Basin (the Basin and Range region) in the western United States (Figure 61).

How Continents Form

Accretions of terranes. The development of a series of mountain belts along a continent's margins increases the size of the continent by adding new continental crust (accretion). In most cases, a continent consists of an older core (craton) surrounded by progressively younger rocks. Mountain ranges are sometimes called **tectonostratigraphic terranes,** or just **terranes,** which represent regions of geologic continuity distinct from neighboring mountain ranges. Terranes can range up to thousands of square kilometers in area. **Accreted terranes** are those that appear to have formed in place along a continent's margin through accumulation and orogeny. A **suspect terrane** is one that does not fit the regional pattern or has conflicting age dates; an **exotic terrane** is one that did not form naturally through accretion and has likely collided with the continental margin. Exotic terranes have distinctive rock types, metamorphic and structural histories, and ages of formation. Paleomagnetic data can sometimes be used to reconstruct an exotic terrane's path of migration. Such terranes can be island arcs, microcontinents such as New Zealand, or rifted fragments of distant continents. North America is composed of over fifty distinct geologic terranes; twelve of these have been accreted to western North America during the past 200 million years.

The earth is estimated to be about 4.5 billion years old. Our knowledge of its history comes from a number of sources. The geologic time scale is constructed through scientific methods and calculations as well as from the interrelationships of geological features as observed in the field. The **principle of uniformitarianism** ("the present is the key to the past") is helpful in that we can accurately measure the rates of geologic processes we see today and apply them to the geologic past. For example, we know layers of sediment build up on the ocean floor at the rate of about 1 millimeter per year. Thus, it would take over one million years of sedimentation to form a unit of shale 1,000 meters thick.

Geologists recognize two different kinds of time: relative time and absolute age. **Relative time** concerns the sequence of geologic events, and **absolute age** measurements concern the actual age of a rock or mineral.

Relative Time

Basic principles. As previously described in this book, geologists use some basic, simple principles to unravel "which came first":

- The **law of superposition** states that in an undisturbed sequence of sedimentary rocks or lava flows the overlying rock is younger than the underlying rock.

- The **law of original horizontality** states that most sedimentary rocks (an exception is cross-bedded sediment) formed as nearly horizontal layers. Any layered sequences that are now tilted were moved by later geologic processes.

- Any rock that cross-cuts another rock is younger than the rock it cross-cuts. This rule applies also to mass wasting and erosion; whatever is eroded had to exist prior to the beginning of erosion.

- The **law of faunal succession** states that fossil species succeed one another in undisturbed rocks in a definite and recognizable order around the world.

- If fragments of one rock type are observed as inclusions within another rock type, the first rock type had to exist prior to the rock type that hosts its inclusions.

- Unconformities are erosional surfaces that represent gaps in geologic time between the formation of the lower rock surface and the overlying sedimentary or volcanic layers.

If one applies these principles to Figure 63, the sequence of geologic events, from oldest to youngest, is as follows:

1. The sequential deposition of layers A through J

2. Tilting and erosion of these units

3. Deposition of formation K, creating an unconformity

4. Intrusion of granite

5. Deposition of formation L

6. Intrusion of cross-cutting dike

7. Erosion of formation L, followed by the deposition of formation M on the nonconformity

8. Subsequent deposition of formations N and O

9. Surface erosion, resulting in formation of drainage patterns

Sequence of Geologic Events

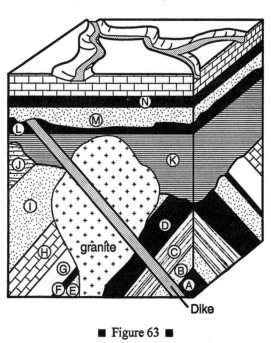

granite

Dike

■ Figure 63 ■

Geologic Correlations

Physical continuity. Because the geologic record is rarely complete or fully exposed, geologists are not always able to physically trace a rock formation or contact over hundreds of kilometers. A continuous, visible contact is the **physical continuity** that is the easiest way to prove that rocks in two different areas are the same. More often both sequences of rocks need to be compared in detail before it can be concluded that a particular rock type in both areas is the same formation.

This examination includes looking at different rock units above and below the formation in question. For example, if you are studying a red shale/sandstone/green shale/limestone/chert sequence and find the same sequence 20 kilometers away, you could assume the green shale is part of the same unit and formed at the same time under similar conditions. The geologic history of an area can be reconstructed by correlating rock exposures within the area. Outcrops may show only parts of a sedimentary sequence, but by comparing the exposed sections, one may identify distinctive **marker beds** in them that may allow the complete reconstruction of the sequence.

The fossil record. Geologic time periods can be well defined by the **fossil** remains of animal and plant species within them. **Paleontology** is the study of fossils. Rocks that contain the same kinds of fossils formed over the same range of geologic time in which the species existed on the earth. **Index fossils** are those species that lived only during a restricted period and that identify the narrow time range during which the host rock could have formed. **Fossil assemblages** are groups of different kinds of fossils that coexisted and are more useful than single fossils in determining the age of the formation and the environment of deposition. Correlations of fossils and fossil assemblages have allowed geologists to identify rock units that formed during the same span of geologic time all over the world.

The geologic time scale. Using fossil evidence, geologists developed the **standard geologic time scale,** which divides the earth's age into eons, eras, periods, and epochs (Table 3). When the time scale was first developed, the earliest forms of fossil life were thought to be no older than the Cambrian period, which began about 570 million years ago. The length of time the earth existed prior to the Cambrian (from about 570 million years ago to 4.5 billion years ago) is called the Precambrian. The Precambrian is further subdivided into categories based on age dates and tectonic features.

Absolute Age

The standard geologic time scale was devised according to relative time relationships observed in rocks across the world. Determining the actual ages of these time spans, and thus establishing the beginning and ending dates of geologic eons, eras, periods, and epochs, became possible with the discovery of radioactivity. The important boundary between the Paleozoic era and the Precambrian era is dated at about 570 million years ago; the Mesozoic era (the "Age of the Dinosaurs") started about 245 million years ago and ended 66 million years ago.

Radioactive elements decay at known rates of speed. This radioactive decay begins after the elements are locked into crystalline mineral structures. Some elements have variations called **isotopes,** which are atoms that contain different numbers of neutrons in their nuclei. For example, uranium has the isotopes U-235 and U-238; U-238 has three more neutrons than does U-235.

Radioactive decay is the breakdown of isotopes that contain unstable nuclei. As an element decays it creates a series of **daughter products.** For example, uranium-238 loses protons and neutrons during its decay, going through a series of intermediate daughter products to form its end product lead-206, a stable isotope. The rate of radioactive decay is constant. By determining the relative amounts of a radioactive isotope and its decay products in a mineral, the *age* of the mineral can be determined. Other decay reactions that are used to calculate absolute age are carbon-14 to nitrogen-14, potassium-40 to argon-40, rubidium-87 to strontium-87, thorium-232 to lead-208, and uranium-235 to lead-207.

An isotope's **half-life** is the time it takes for half of a known quantity of radioactive material to convert to its daughter product. For example, the half-life of U-238 is 4.5 billion years. Thus, if you began with one gram of U-238, 4.5 billion years later only one-half gram would remain. After another 4.5 billion years, only one-quarter of the original amount would remain.

Depending on the kind of isotope being analyzed, isotopic ratios are measured in single mineral crystals or larger pieces of rock.

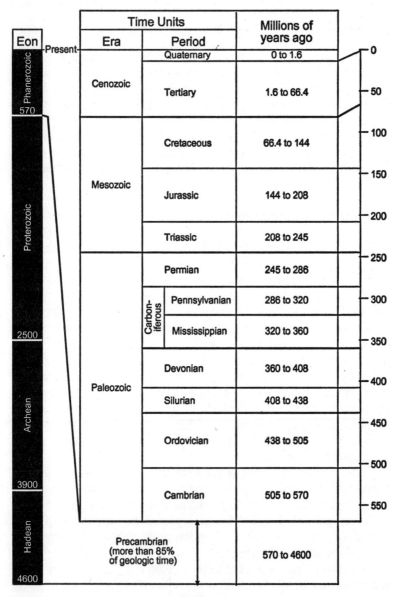

Eon		Time Units		Millions of years ago
		Era	Period	
Phanerozoic	Present	Cenozoic	Quaternary	0 to 1.6
			Tertiary	1.6 to 66.4
570		Mesozoic	Cretaceous	66.4 to 144
Proterozoic			Jurassic	144 to 208
			Triassic	208 to 245
		Paleozoic	Permian	245 to 286
2500			Carbon-iferous Pennsylvanian	286 to 320
			Carbon-iferous Mississippian	320 to 360
Archean			Devonian	360 to 408
			Silurian	408 to 438
			Ordovician	438 to 505
3900			Cambrian	505 to 570
Hadean		Precambrian (more than 85% of geologic time)		570 to 4600
4600				

Scale (Millions of years ago): 0, 50, 100, 150, 200, 250, 300, 350, 400, 450, 500, 550

■ Table 3 ■

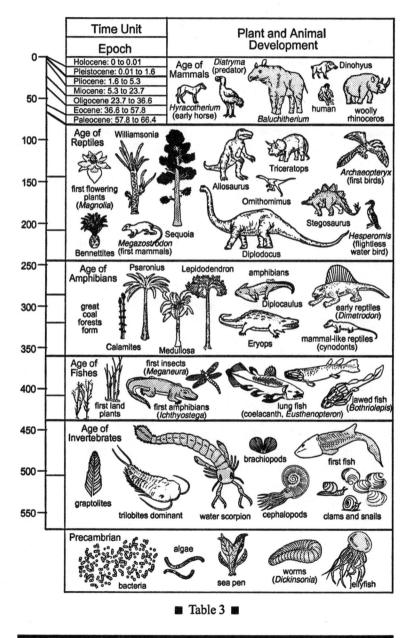

Time Unit	Plant and Animal
Epoch	Development

Holocene: 0 to 0.01
Pleistocene: 0.01 to 1.6
Pliocene: 1.6 to 5.3
Miocene: 5.3 to 23.7
Oligocene 23.7 to 36.6
Eocene: 36.6 to 57.8
Paleocene: 57.8 to 66.4

Age of Mammals — *Diatryma* (predator), Dinohyus, *Hyracotherium* (early horse), *Baluchitherium*, human, woolly rhinoceros

Age of Reptiles — Williamsonia, first flowering plants (*Magnolia*), Bennettites, *Megazostrodon* (first mammals), Sequoia, Triceratops, Allosaurus, Ornithomimus, Diplodocus, Stegosaurus, *Archaeopteryx* (first birds), *Hesperornis* (flightless water bird)

Age of Amphibians — great coal forests form, Psaronius, Calamites, Lepidodendron, Medullosa, amphibians, Diplocaulus, Eryops, early reptiles (*Dimetrodon*), mammal-like reptiles (cynodonts)

Age of Fishes — first land plants, first insects (*Meganeura*), first amphibians (*Ichthyostega*), lung fish (coelacanth, *Eusthenopteron*), jawed fish (*Bothriolepis*)

Age of Invertebrates — graptolites, trilobites dominant, water scorpion, brachiopods, cephalopods, clams and snails, first fish

Precambrian — bacteria, algae, sea pen, worms (*Dickinsonia*), jellyfish

■ Table 3 ■

Radiometric age dating works best on igneous, volcanic, or metamorphic rocks. It is important to select "fresh" rock that has not been chemically or structurally altered by deformation, weathering, hydrothermal alteration, or metamorphism. The calculated age is the age from which the rock or mineral stabilized. The interpretation of the age date also depends somewhat on the kind of rock being analyzed. For example, age dates from volcanic rocks that cooled quickly can give an age that is very close to the age of the eruption of the volcanic flow. Since it takes plutonic rocks millions of years to cool, the age from a mineral that formed in the pluton could be close to the age of intrusion or close to the age of the final crystallization, depending on when it formed in the intrusion. Minerals that could generate original age dates from rocks that have been metamorphosed are usually destroyed by the metamorphism; ages derived from metamorphic rocks typically indicate the time the rocks cooled and stabilized after metamorphism. Dates can usually be determined to within a few million years; in the scope of geologic time, that is a relatively small margin of error.

Absolute age dates have confirmed the basic principles of relative time—for example, a uranium-lead date from a dike that intrudes into an older rock always yields an absolute age date that is younger than the absolute age date of the enclosing rock. Thousands of comparisons between absolute ages and relative time relationships have proven that radiometric age dating works.

Geologists have calculated the beginning and ending dates for the eons, eras, periods, and epochs of the geologic time scale, and major geologic events. To date, the oldest rocks discovered on the earth are from northwestern Canada and are 3.964 billion years old. The development of complex life in the Cambrian period occurred about 570 million years ago, or only in the last 13 percent of the earth's vast history. Age dating has permitted the age specificity of the internationally accepted geologic time scale shown in Table 3.

A Summary of Earth's History

The Precambrian. The vast unit of time known as the **Precambrian** started with the origin of the earth about 4.5 billion years ago and ended 570 million years ago. Largely thought to be a hot, steaming, and forbidding landscape, the primitive crust of the newly condensed planet continued to cool. The crust consisted largely of igneous intrusions and volcanic rocks, and sediments that were eroded from this irregular surface. Geologic remnants from this time are the highly deformed and metamorphosed cratons of the continents. The Precambrian is subdivided, from oldest to youngest, into three eons, the **Hadean** (4600–3900 million years ago), **Archean** (3900–2500 million years ago), and **Proterozoic** (2500–570 million years ago). Little is known about the Hadean because there are so few rocks of that age, and those that do exist are intensely deformed and metamorphosed. The Archean was dominated by crustal building and the development of extensive volcanic belts, arcs, and sedimentary basins that were probably related to plate tectonic activity. Marine rocks including chert contain the fossil remains of microscopic algae and bacteria. The Proterozoic is known for large-scale rifting of continental crust across the world and the filling of these rifts with huge amounts of sedimentary and volcanic rocks. Extensive iron deposits formed in shallow Proterozoic seas, indicating there was enough free oxygen to precipitate iron oxide minerals (for example, hematite [Fe_2O_3]) from the iron in the water. The increase in the amount of free oxygen is thought to be a result of photosynthetic action by primitive life forms in the sea. The fossil record has preserved layered algal mounds called stromatolites, an abundance of microscopic species, and trails and burrows from wormlike organisms.

The Paleozoic era. The **Paleozoic era** (570–245 million years ago) was long believed by geologists to mark the beginning of life, because of the sudden abundance of complex organisms with hard parts in the fossil record. These organisms included trilobites and shelled animals called cephalopods (cephalopods were the ancestors of modern squids and octopi). Life was restricted to the sea and included graptolites, brachiopods, bryozoans, and mollusks.

A single southern landmass consisted of what is today South America, Africa, India, Antarctica, and Australia. In the northern hemisphere, land masses that represent North America, Siberia, northern Europe, western Asia, and China had not yet joined the southern landmass. North America was essentially a lowland that was periodically flooded by the ocean, forming extensive deposits of sandstone, limestone, and barrier reefs.

By the end of the Paleozoic, all of the continents had come together to form Pangaea. This formation resulted in extreme seasonal weather conditions and one of the greatest periods of extinction in the earth's history—up to 75 percent of amphibian species and 80 percent of marine species disappeared. This time was also marked by the rapid development of land plants, forests of short trees, armor-plated fishes, sharks, and bony fishes. The Devonian period, the fourth period in the Paleozoic era, is known as the "Age of the Fishes." Air-breathing amphibians began to move from the ocean to land. Large tropical swamps dominated much of the landscape.

The Mesozoic era. The **Mesozoic era** occurred from about 245 million to 66 million years ago. The fossil record from this era (the "Age of the Dinosaurs") is dominated by a multitude of dinosaur species. Common sedimentary deposits are red sandstones and mudstones. The low-lying areas were frequently flooded by shallow marine transgressions. Tropical conditions resulted in extensive swamps that later became coal beds. By the mid-Mesozoic, Pangaea rifted into northern Laurasia and southern Gondwanaland. Igneous and volcanic activity formed the mountain ranges in western North America.

In the Mesozoic era, new trees such as conifers and ginkgoes appeared. Reptiles laid eggs on land. Dinosaur species included meateaters, herbivores, winged reptiles, and marine reptiles. Mammals were just beginning to emerge during this time. The end of the Mesozoic is marked by more mass extinctions, especially of the dinosaurs.

Surviving species included turtles, snakes, crocodiles, and various lizards.

The Cenozoic era. The **Cenozoic era,** also called the "Age of Recent Life" or "Age of Mammals," encompasses the last 66 million years of the earth's history. Life forms continued to become more complex. The Cenozoic has the most complete geologic record of any era because it is so recent. The continents were fully separated. Plate tectonic activity created many orogenic and volcanic events in North America, including the western fault-block mountains and huge lava flows. Eastern North America was tectonically stable, and the Appalachians eroded to lower elevations. Valleys in the western part of the continent were filled with great thicknesses of sediments from the mountain ranges.

The fossil record indicates a diverse array of mammals (including marsupials and placentals), flowering plants, grasses, and microscopic foraminifera. New birds and mammals evolved that were adapted to the new vegetation species. Prehistoric humans also began to emerge. Waves of mass extinctions occurred toward the end of Pleistocene epoch, including those of mammoths, mastodons, saber-toothed cats, ground sloths, and camels. North America underwent multiple glaciations in the last 20,000 years, which helped mold the landscapes we see today.

Early humans began to recognize the importance of the earth's **natural resources** first for shelter, weapons, and tools. Metal such as copper was mined, smelted, and used for trading with other civilizations. We have now developed thousands of uses for our natural resources, uses that support our modern-day standard of living. Most of our lifestyles depend on nonrenewable resources that come from the earth—resources such as petroleum, coal, metal, sand, and gravel. Our demand for these materials far exceeds the amounts recycled. Thus, we have a constant need to discover new deposits, which are becoming harder to find as the "easy" ones are discovered at the surface and mined out. Geologists and scientists are beginning to look for deposits deeper below the surface, a more challenging and expensive endeavor. Geophysical and satellite photography techniques are being refined to allow us to "look" below the surface of the earth. New metallurgical and processing technologies permit the mining of what were once considered unmineable deposits.

Resources and Reserves

A **resource** is that amount of a geologic commodity that exists in both discovered and undiscovered deposits—by definition, then, a "best guess." **Reserves** are that subgroup of a resource that have been discovered, have a known size, and can be extracted at a profit. For example, of the world's estimated oil resource of three trillion barrels, the world's *reserves* are estimated at about a third of that amount. Factors that affect profitability include the demand, market price, mining costs, transportation costs, new technologies that can extract the material at a lower price, taxes, environmental laws, and government price controls. Understandably, a deposit's economic value can change with time as these factors change.

Metallic Deposits

Metals occur in all kinds of rocks but usually in concentrations that are too low to be mined. **Metallic ore deposits,** however, are relatively rare concentrations of metal-bearing minerals (usually sulfides) that contain enough metal to be profitably mined. Again, the profit line is dependent on a number of economic factors. Our most important metals are iron, copper, aluminum, lead, zinc, silver, gold, chromium, nickel, cobalt, manganese, molybdenum, tungsten, vanadium, tin, mercury, magnesium, platinum, and titanium.

Mineral exploration is the practice of exploring for and discovering new ore deposits. Exploration is becoming progressively more challenging as the ore deposits exposed at the surface are discovered and mined. Future exploration will focus on developing techniques that will help find ore deposits that are hundreds or thousands of feet below the surface and impossible to detect at the surface.

Metallic ores occur in every kind of rock and some varieties of soil. The metallic minerals are concentrated into rich masses by igneous, hydrothermal, or erosional/weathering processes. Metals such as chromium, platinum, nickel, copper, and iron can precipitate as sulfide minerals in a cooling body of magma. **Magmatic deposits** result when the minerals settle to the bottom of the intrusive body and form thin, high-grade layers. **Hydrothermal deposits** rich in copper, lead, zinc, gold, silver, molybdenum, tin, mercury, and cobalt form from hot solutions that circulate through fractured country rock. The solutions come from nearby intrusions or heated meteoric water. Much of the dissolved metal in the solutions is leached from the surrounding rocks through which the solutions migrate. Changing pressures and temperatures precipitate the metals as sulfides or pure metal, such as gold, silver, and copper. This process is usually repeated many times until the heat source has cooled or the fracture systems have become filled with mineral deposits.

Common types of hydrothermal deposits are contact metamorphic, hydrothermal, disseminated, and hot springs deposits. **Contact metamorphic deposits** result from hot solutions that migrate from a cooling intrusion and deposit minerals in cracks in the surrounding country rock. **Hydrothermal veins** are also mineral deposits in faults and cracks but are not necessarily related to an intrusive body. The fluid can be meteoric water that has moved downward toward a heat source, been heated, and ascended, leaching metals along its path. The sulfides are later deposited a considerable distance from the heat source. Some of the richest gold and silver deposits in the world are hydrothermal veins. **Disseminated deposits** are those in which the metal is evenly distributed in generally low concentrations throughout large masses of rock. An important type of disseminated deposit is the **porphyry copper deposit,** in which copper and molybdenum are found in porphyritic intrusive rocks. Huge, low-grade, multimillion-ounce disseminated gold deposits have been found in sedimentary rocks in Nevada. **Hot springs deposits** are minerals that formed in response to hot spring activity at the surface of the earth. These can be rich in gold, silver, antimony, arsenic, and mercury.

Ore deposits can form also by other processes at the earth's surface. **Mississippi Valley-type deposits** are concentrations of lead and zinc that are thought to be deposited in porous limestones and sandstones by low-temperature water that was driven out of deeper sediments by compaction. These deposits are common in the central United States over relatively stable crust and may be one of the few deposit types not related to some kind of igneous heat source. The ore minerals in most of the world's iron and manganese reserves were chemically precipitated in the ocean and accumulated on the sea floor. **Placer deposits** are heavy metallic minerals, such as iron or titanium minerals, or native gold or diamonds, that have been concentrated by wave or water action in a river or beach environment. The source of the minerals may be far upstream and contain very low amounts of these minerals. The weathering, erosion, downstream transport, and deposition result in concentrations of the minerals that can be profitably mined. **Lateritic weathering** results in residual deposits that became enriched through the chemical breakdown and removal of most of the elements of the rock. For example, in tropical climates, nickel and aluminum are left behind as their host rocks are chemically weathered, forming enriched, high-grade **supergene deposits** that can be mined. High-grade supergene gold and copper deposits can form also when a low-grade deposit is weathered downward, and the metal accumulates in place. A rusty, iron-bearing cap called a **gossan** is often the only remnant of a weathered metallic ore deposit at the surface. Finding a gossan may indicate additional minerals exist below the zone of weathering.

Most metallic ore deposits are a result of plate tectonic activity. High heat flows and convection currents at divergent plate boundaries, such as midoceanic ridges, create submarine hot springs called **black smokers** that deposit solid masses of metallic minerals. These are important environments for the deposition of iron, copper, zinc, lead, gold, and silver. Metal-laden solutions that are denser than water can collect in basins on the ocean floor, forming rich deposits. Island arc systems that develop along converging boundaries also create massive sulfide deposits rich in base metals, as well as hot-spring gold-silver deposits on the flanks of andesitic volcanoes. The development of metallic ores in intrusive rocks, such as porphyry copper deposits, is related to the partial melting and rising of crustal material along subduction zones. Metal in these systems may also be contributed from sea-floor deposits that were subducted and became part of the new magma. Chromite may occur also in ultramafic intrusions in the new oceanic crust that forms at divergent boundaries.

Energy Resources

Our energy resources include petroleum and petroleum products, coal, uranium (nuclear reactions), and geothermal resources. At present, about 90 percent of the energy needs of the United States are supplied by coal, oil, and natural gas. Petroleum is important also in the production of plastics, asphalt, and thousands of related products.

Fossil fuels. **Fossil fuels** are oil, natural gas, and coal. The general term **petroleum** includes both natural gas and crude oil. **Crude oil** is a liquid containing hydrocarbons (molecules made from only hydrogen and carbon) that forms in organic- or fossil-rich sediments and rocks. The hydrogen and carbon in the oil comes from the breakdown of the organic material over time. **Natural gas** is a gas that contains hydrocarbons and that usually occurs with crude oil.

Petroleum forms in marine sedimentary rocks that contain abundant organic remains from microscopic organisms such as algae. Continental shelves contain basins that capture thick accumulations of organic debris. This material lithifies into a **source rock** that is buried by overlying sediments, and the resulting increased pressure and temperature conditions convert the organic material into hydrocarbons.

In response to the confining pressure, petroleum moves outward and upward along zones of increased permeability into a **reservoir rock**. Reservoir rock, such as sandstone or limestone, has the high porosity and permeability necessary to hold large accumulations of petroleum. The petroleum migrates into a **trap** (either structural or stratigraphic) in the reservoir rock. **Structural traps** include faults between permeable and impermeable rocks, thrust faults, and folds such as anticlines. For example, petroleum will collect in a porous limestone reef below the contact with an overlying, impermeable unit such as shale, forming a pocket. A **salt dome** results when a bed of rock salt is under pressure; the salt extends upward plastically through a sedimentary sequence, disrupting the sediments and creating open spaces that trap petroleum. **Stratigraphic traps** are naturally occurring changes in a sedimentary sequence that trap migrating oil and gas, such as a porous reef structure in a limestone unit. A sandstone unit surrounded by shale is another stratigraphic trap. The occurrence of **oil pools** in a number of traps in one area is called an **oil field.**

Crude oil and gas are recovered from oil fields through a series of drilled wells. The petroleum may rise to the surface through the well as a result of its high confining pressure, or it may need to be pumped. Also, water or steam can be pumped into the oil pool from the surface to increase the pressure on the oil and its viscosity. The oil is shipped to a refinery and separated into natural gas, gasoline, kerosene, other oils, and asphalt. A huge variety of petrochemicals produced from petroleum are used in nearly every manufactured product we depend on today, including plastics and synthetic rubber. As the more easily discovered oil fields are pumped dry, oil companies have moved offshore to undertake risky and high-cost exploration drilling along continental shelves across the world.

With only about 5 percent of the world's population, the United States annually consumes over a quarter of the world's total oil production. At present, the United States has only a twenty-five-year supply of oil remaining and imports nearly half of the oil it uses. Similarly, natural gas reserves in the United States are expected to be depleted within thirty-five years. Future sources for natural gas will include gas trapped in coal beds.

Oil sands (tar sands) are sandstone deposits that have been cemented with tar or asphalt (blackish, solidified residues from petroleum). Famous deposits include those of the La Brea Tar Pits in Los Angeles, California, and the Athabasca Oil Sand in Alberta, Canada. Oil sands are strip-mined and processed. Venezuela also has large reserves of oil sand. **Heavy crude** is a dense, viscous petroleum that flows so slowly it is usually left behind in an oil field. Efforts are continuing to extract this material, including pumping in steam or other solvents to make the crude less viscous.

Oil shales are organic-rich shale formations from which oil can be extracted. The shale formed from muds on the bottom of large shallow lakes. Oil shales tend to be low-grade and difficult and costly to mine. New technologies are being used to explore ways to extract the oil from the rock in place, including heating the rock with microwaves to separate the oil. The United States has large oil shale resources in Montana, Utah, Colorado, and Wyoming, but at present they are not feasible to mine. Oil shales will ultimately be exploited when the cost of finding new oil fields gets too high.

Coal is a dark-colored sedimentary rock that contains a high percentage of organic plant material. Coal is representative of warm, lush, swampy environments and usually contains abundant plant fossils. Different kinds of coal result from different degrees of compaction and depth of burial. **Peat** is unlithified organic material that is solid enough to be cut into blocks and burned for fuel. Burial and increasing pressure and temperature convert peat into a soft, brown coal called **lignite.** Continued pressure results in **subbituminous coal** and **bituminous coal,** which are soft, black, banded, and sooty.

Metamorphism converts these varieties of coal into **anthracite, a** hard, black, shiny form of coal that is dust free. Coal beds, or **seams,** range in thickness from a few centimeters to nearly 30 meters. Coal is mined using underground, open-pit, and strip-mine methods. The United States has an impressive coal resource of nearly four trillion tons and consumes or exports about a billion tons a year. Most of the coal in the United States is produced in Kentucky, West Virginia, and Pennsylvania.

Coal once provided nearly all of the United States' energy needs. That figure has dropped to less than 25 percent because of the abundance of petroleum, oil, and natural gas and the negative environmental effects of burning coal. Coal is an important ingredient in manufacturing steel. Oil and gas can also be produced from coal. New, cleaner ways of using coal are being researched as the country's petroleum reserves are being depleted.

Uranium and geothermal sources. Uranium is used to generate nuclear power. It is found in the minerals pitchblende and carnotite, which are mined from sandstone deposits in the western United States and Canada. Lower-grade uranium also occurs in organic black shales and phosphate deposits. Nuclear generators are used to produce electricity. Nuclear power supplies about 8 percent of the United States' energy needs.

Geothermal sources can also generate electricity but represent less than 1 percent of the energy consumption in the United States. More geothermal power may be used in the future, especially if deeper heat sources across the nation can be located and exploited.

Renewable resources. Research is continuing on developing energy from **renewable resources**—power from sources such as wind, sun, running water, waves, and ocean-currents and from burning hydrogen from the breakdown of water. The obvious attraction is that these resources are renewable, and they will become more important as our supplies of nonrenewable resources continue to dwindle.

Nonmetallic Resources

Although not as visually impressive as glittering silver, gold, or sulfides, nonmetallic resources are just as important to our economy. They are more abundant and less expensive than metallic ore deposits. **Stone** is removed from quarries as blocks for building purposes or is crushed for highway construction. Most stone is limestone or granite. Limestone is also an important component of cement and is used to replenish lime in farm soil. **Sand** and **gravel** are mined extensively from shallow open pits for highway construction, cement, and concrete. Well-sorted sand dune, beach, and glacial outwash deposits are the best sources for sand and gravel. Glass is also produced from **silica sand. Bentonite,** a clay mineral, is known for its ability to absorb water and is used in cement and cat litter. Volcanic material such as that in cinder cones can also be a gravel source. Finely ground volcanic pumice is used as an abrasive.

Evaporitic rocks and phosphate deposits, enriched in phosphorous, nitrogen, and potassium, are important sources of agricultural fertilizers. **Gypsum,** another evaporitic mineral, is important in the manufacture of plaster and wallboard. **Rock salt** is mined from evaporitic salt beds and salt domes for use as a deicer and a food preservative. **Sulfur** is recovered from salt domes to make fertilizers, acids, and explosives.

Graphite is widely applied as a lubricant. **Asbestos** is a unique variety of the mineral serpentine that can be woven into fireproof material (although it is rarely used for this purpose today because it is a carcinogen). **Talc** is a fine-grained mineral substance that is used as a lubricating powder. **Borates** are an important component in cleaning compounds. **Barite** is a very heavy mineral that is used as a drilling additive in the oil industry. **Fluorite** is used in toothpaste, Teflon, and the steel industry.

In addition to their value as jewelry, **gemstones** have important industrial applications. Gemstones include diamonds, rubies, emeralds, beryl, garnet, topaz, and zircon. Gemstones are frequently used as abrasives because they are hard minerals. For example, diamonds are used in drill bits and saws designed to cut through rock and steel.

Diamonds, the hardest substance on the Mohs hardness scale, can also be manufactured in the laboratory for industrial use.

Recycling and Conservation

Our geologic resources are nonrenewable. The more we consume, the less will be available for future generations. Even though new technologies are providing ways of using other materials in industry, recycling will continue to play an increasingly important role in the years to come. The amount of consumption of resources continues to vastly exceed the amount of the recovery and reuse of those resources through recycling. Conservation efforts target using less of a commodity, such as petroleum, to slow down the depletion of known reserves. How quickly recycling and conservation efforts increase in the future will likely be directly related to the increasing cost and demand for certain commodities.

Advances in photography, astronomy, telescopes, analytical techniques, telecommunications, and space flight have given us more detailed glimpses of our solar system. Space probes have measured the chemical compositions of the atmospheres on planets, and rock samples have been recovered. We have mapped the surfaces of our Moon and nearest planets in detail. Geologic features similar to those on Earth have been identified. The theory of uniformitarianism can be applied to outer space as well as to Earth. For example, the braided patterns on the surface of Mars were probably formed by running water as they are here.

Scientists believe that our solar system—consisting of the Sun, nine planets, and numerous moons—formed about the same time, 4.5 billion years ago. The four planets closest to the Sun—Mercury, Venus, Earth, and Mars—are **terrestrial planets.** They are called terrestrial because their densities (of 3 g/cm^3 or more) are similar to Earth's. The rest are called **jovian planets** and have densities below 2 g/cm^3. All nine planets have been studied by uncrewed spacecraft. Measurements have indicated that all the planets appear to have solid cores.

The kinds of tectonic, magmatic, and surficial processes that have shaped Earth's surface have also affected the surfaces of the other planets. Likewise, the impact cratering from meteorites so visible on other planets has also occurred on Earth and has been suggested as a reason for the mass extinction of the dinosaurs.

Mercury

Mercury is the closest planet to the sun. Spacecraft data indicate the planet has virtually no atmosphere and is densely cratered. Basaltic lava flows have also been identified. A metallic core that accounts for about three-quarters of the planet's mass has been offered as an explanation for the planet's high density of 5.4 g/cm^3. Mercury also has a weak dipolar magnetic field. The lack of any active volcanic, tectonic, or erosional processes suggests Mercury is a "dead" planet.

Venus

Information about **Venus** has come from radio signals transmitted by spacecraft that have landed on the surface. Surface temperatures are about 500 degrees centigrade. The atmosphere is composed of carbon dioxide, which traps much of the Sun's radiation.

Broad plains dotted with thousands of low-viscosity basaltic shield volcanoes cover about 80 percent of the surface. Steep-sided volcanic structures resembling felsic lava domes on Earth suggest that felsic magmas may occur. Rocks analyzed by a Russian spacecraft contained high potassium concentrations, also suggestive of felsic magmatic differentiation.

Radar mapping of the surface has identified elongate mountain ranges that seem to have been formed by extension and compression, possibly as a result of plate tectonic activity or mantle convection and mantle plumes.

Earth's Moon

Our knowledge of Earth's **Moon** has come from years of detailed telescopic study and uncrewed and crewed space flight. Seismic detectors left by astronauts record about four hundred weak moonquakes a year. The seismic data indicate the Moon's interior is layered underneath a 65-kilometer-thick crust made up of basalt and feldspar-rich rock. Mantle material is inferred to exist beneath the crust. Most scientists think the moonquakes are related to the gravitational pull of the earth, which causes small crustal movements along faults. Seismic calculations indicate the Moon is solid down to a depth of about 1,000 kilometers, the lower limit of the Moon's lithosphere (crust and rigid upper mantle).

The large, smooth lowlands on the lunar surface are regions covered by basaltic lava flows. Geophysical measurements suggest the mountainous regions are isostatically balanced, indicating that at one time the Moon's lithosphere was plastic enough to let the thicker crust "float." Rock samples retrieved by crewed expeditions consist of feldspar-rich anorthosite, an igneous rock, and basalt. Age dates vary from about 3.0 to 4.5 billion years—the proposed beginning of the solar system.

Mars

Mars is similar to Earth in that it has about a twenty-four-hour day, an atmosphere, polar ice caps (carbon dioxide), and winter and summer seasons. Unlike Earth, Mars does not have a dipolar magnetic field. To date, no evidence of life has been discovered on the planet.

The northern hemisphere is marked by shield volcanoes and volcanic cones. Craters are widely scattered. These volcanic mountains are three times as high as the tallest volcanic mountains on Earth. Olympus Mons is the largest volcanic structure discovered so far in the solar system. The tremendous size of the volcanoes suggests that the magma sources supplied the volcanic vent for a very long time and that the lithosphere is very strong. Horst and graben fault basins have also been identified. The volcanic activity is thought to be fairly recent, since many of the cones have not been pitted by meteoric impacts (an application of relative time principles). Craters are also rare on the volcano slopes, suggesting the most recent layers of volcanic flows are less than 100 million years old. In contrast, the southern hemisphere is studded with thousands of impact craters and is not as volcanically active.

The Martian surface, weathered and composed of clay and sulfate materials, is shaped by winds that form sand dunes. Meandering braided structures look as though they were formed by running water or gigantic floods. Flooding could occur if the frozen surface were suddenly melted by magmatic activity or a change in climate.

Jupiter and Saturn

Jupiter and **Saturn** are the best-studied of the jovian planets. It is from these huge, gassy planets that we have learned most about the primordial gases from which all of the planets are believed to have condensed. Jupiter's atmosphere consists mostly of hydrogen, ammonia, helium, and methane. Jupiter and Saturn give off twice as much energy as they receive from the Sun, which suggests they are still undergoing the process of gravitational contraction and condensation. One of Jupiter's moons, Io, is volcanically active, and its bright orange colors may indicate sulfur-rich rocks. Photography has revealed long lava flows radiating out from the volcanic cones. Abundant andesitic volcanism has created lava plains and huge shield volcanoes. Few impact craters are seen because they are quickly covered by volcanic flows. Another kind of volcanic activity throws explosive sprays of sulfurous material up to 300 kilometers into the atmosphere. Jupiter's outer moons—Europa, Ganymede, and Callisto—are covered in ice. Intriguing patterns of grooves, striations, and cracks may be the result of some kind of plate tectonic activity that is affecting their frozen, rigid surfaces. Saturn's moons are also ice covered and pitted with craters. The moon Titan has an atmosphere that consists mostly of nitrogen, acetylene, ethane, ethylene, and hydrogen cyanide. It is conceivable that Titan's surface is covered by continents of ice and by liquid ethane and methane.

Much of what we know about our solar system is through the application of uniformitarianism—that the "present is the key to the past" and thus that the geologic processes we see today were active in the geologic past. The integration of our ongoing exploration of Earth and the planets around us will lead us to a better understanding of how our solar system formed. Our heightened awareness of geology, its active processes, the resources we consume, and the end products we generate can lead us to a better coexistence with the planet Earth.

aa flow a lava flow that develops a partially solidified surface as it moves forward, which breaks the surface into a rough, rubbly mass.

ablation the loss of ice and snow from a glacier, generally through melting or evaporation.

abrasion of a stream, the process by which a stream's rocky bed is eroded away by the constant friction and impact of tumbling rock fragments, gravel, and sediment.

abyssal fan a fan-shaped accumulation of sediment that forms at the mouth of submarine canyons.

abyssal plain a very flat expanse of horizontally deposited sediment that accumulates on the ocean floor at the base of a continental rise.

accreted terrane a terrane that appears to have formed in place along a continent's margin through accumulation and orogeny.

accretionary wedge an accumulation of marine sediment, derived from the subducting plate, that builds up at the edge of a subduction zone.

acid rain an environmentally harmful acidic rain that results from rain mixing with chemical pollutants in the atmosphere.

active continental margin marked by a landward continental shelf, a continental slope that forms a sidewall of an oceanic trench and is much steeper than that of a passive continental margin, and an irregular ocean bottom that may contain volcanic seamounts; an area of earthquake and volcano activity.

advancing glacier a glacier that exhibits outward or downslope movement.

aftershock one of the small earthquakes that may follow the main earthquake.

A horizon the uppermost soil horizon; characterized by the downward movement of water.

alkali soil a pedocal type of soil toxic to plant growth because of its high salt content.

alluvial fan a feature similar to a delta; a large, fanlike accumulation of sediment dropped where a stream emerges from rugged terrain, such as the edge between a mountain canyon and a flat plain.

alpine referring to high mountain regions.

alpine glaciation glaciation usually restricted to deep valleys in high mountainous terrain.

angle of repose the steepest angle at which loose material will remain in place.

angular unconformity the contact that separates a younger, gently dipping rock unit from older underlying rocks that are tilted or deformed layered rock.

anticline a fold that is arched upward to form a ridge.

aphanitic fine grained.

aquiclude an impermeable layer such as clay that retards the flow of groundwater.

aquifer a porous, permeable, saturated formation of rock or soil through which groundwater flows easily.

aquitard a formation such as shale, clay, or unfractured igneous rocks that retards water flow.

arc-continent convergence the result when intervening ocean is destroyed by subduction, welding an island arc to the continental edge.

arete a sharp ridge that commonly extends downward from a horn to separate two adjacent glacial valleys.

arroyo a narrow gorge with steep walls and a gravel bottom; produced over time by flash floods.

artesian well a well that taps water from a confined aquifer.

aseismic ridge a chain of seamounts and guyots; so called because it is not associated with earthquakes.

assimilation the process by which pieces of the country rock melt and mix within a body of magma.

asthenosphere an area composed of the flexible mantle beneath the lithosphere.

atoll a circular reef in deep water that shelters a lagoon; the result of reef development around the flank of a volcano that has since subsided but to which the corals are still anchored.

aulacogen see **failed rift.**

aureole see **halo.**

axial plane of a fold, the plane that separates rocks on one side of a fold from those that dip in the opposite direction on the other side.

backarc basin the area on the continental side of an island arc or magmatic arc.

backarc thrust belt the belt of rocks that has been thrust toward the continental interior from the magmatic arc area along low-angle faults.

backswamp a poorly drained and marshy area behind a natural levee.

backwash the water that flows back down the beach into the surf zone.

bajada the joining of alluvial fans at the front of a mountain range in a rolling surface of sediment and gravel.

balanced budget of a glacier, a situation in which there is neither advancement nor recession.

bar an elongate sedimentary accumulation of sand or gravel deposited by current action in a stream or other body of water.

barchan dune a solitary, crescent-shaped dune that forms in areas of sparse sand.

barchanoid dune a variety of dune intermediate between barchan and transverse dunes; barchanoid dunes form scalloped rows of sand perpendicular to the wind.

barrier island a large, elongate mass of sand that parallels a coast and forms an island.

barrier reef an elongate reef that develops offshore parallel to a coastline and is separated from the coastline by a deep lagoon.

basal sliding the movement of a glacier generated by its sliding on a thin film of water resulting from the pressure of the glacier's weight.

Basin and Range topography a series of steep mountain ranges separated by broad valley floors.

batholith a pluton larger than 100 kilometers at the earth's surface; usually granitic and made up of diapirs.

baymouth bar see **spit.**

beach the strip of sand or gravel (more rarely silt) that covers a shoreline from the low-water edge to a well-defined point of higher elevation.

beach drift the zig-zag pattern by which sediment is moved across a beach face by breaking waves.

beach face the side of a beach facing the ocean.

bedding the pattern (usually horizontal) of layering in which sediments are deposited.

bedding planes demarcations of different layers of sediments.

bed load of a stream, the heavier, coarser-grained earth material that travels on or near the bed of a stream.

bed load of wind, the heavier grains (usually sand) that hop and skip along the ground by saltation.

Benioff zone a zone that slopes downward from an oceanic trench and underneath the overlying crustal plate at 30 to 60 degrees; an area of earthquake origination.

bergschrund a crevasse, commonly filled with rock fall debris, that forms where a glacier separates from a cirque wall.

berm the landward edge of a beach.

B horizon the middle soil horizon into which the leached materials from the A horizon often precipitate.

bituminous coal a common form of coal that is soft and black.

black smoker a submarine hot spring that results from high heat flows and convection currents at divergent plate boundaries and that deposits solid masses of metallic minerals.

blowout a bowl-like depression caused by deflation.

body wave a seismic wave that radiates out from the focus of an earthquake and travels though solid rock.

bottomset bed the finest silt and clay particles that are carried out from a delta into deeper water or slide down a delta front into deeper water.

Bowen's reaction series a description of the progression of mineral formation as magmas cool and crystallize.

braided stream a stream in which the water has lost its main channel and flows in an interconnecting network of rivulets around numerous bars.

breaker a high wave in which the crest falls forward in front of the main body of the wave.

breakwater a wall built parallel to the shoreline to provide quiet water.

breccia rock composed of coarse-grained, angular fragments of broken rocks that have been cemented together and lithified.

brittle strain strain that occurs when a stress is great enough to break or fracture a rock.

budget of a glacier, the ratio between ice gained and ice lost.

butte a landform resulting from the erosion of a mesa.

caldera a depression larger than a crater, at least a kilometer in diameter, that forms at the top of a volcano when the summit is destroyed during an eruption or when the crater floor collapses into the magma chamber below.

caliche a hardpan formed by the precipitation of salt by evaporation.

calving the breaking off of large blocks of ice from a glacier.

capillary action the process by which surface tension causes water to rise up into unfilled pore spaces.

capillary fringe the lower part of the unsaturated zone that draws water upward from the saturated zone.

cast a fossil formed when the organic remains dissolved, leaving an opening (mold) shaped like the organism and later filled with calcite or silica.

cementation the step in lithification in which minerals fill some or all of the pore space and adhere to the sediment fragments, thus producing a sedimentary rock.

chemical sedimentary rock a sedimentary rock resulting from biological or chemical processes, generally underwater, that crystallizes minerals that accumulate on the sea floor.

chemical weathering the process by which rain, water, and atmospheric gases decompose minerals, destroy chemical and mineralogical bonds, and form new minerals.

chill zone the fine-grained edge of a rock intrusion.

C horizon the lowest soil horizon, which lies directly above the bedrock; composed partly of soil and partly of decomposing bedrock fragments.

cinder cone a feature composed of pyroclastic material (not lavas) ejected from a volcanic vent.

circum-Pacific belt an earthquake belt that follows the rim of the Pacific Ocean.

cirque a steep-sided, circular hollow carved in the top of a mountain from an alpine glacier.

clastic sedimentary rock a sedimentary rock formed from the consolidation of material such as gravel, sand, or clay (sediment) derived from the weathering and breakdown of preexisting rocks.

cleavage the ability of a mineral to break along preferred directions, usually along the faces of layered crystals.

coal a dark-colored sedimentary rock that contains a high percentage of organic plant material.

coast the strip of land near the ocean that includes the beach and the immediate inland area beside it.

coastal straightening the process of the headlands being cut back and the flanking beaches being widened.

collision boundary a convergent boundary that separates two continental plates that are pushed into contact.

columnar jointing see **columnar structures.**

columnar structures cooled and contracted flood basalt in vertical, parallel, generally six-sided columns.

compaction the step in lithification in which the grains of sediment are packed more tightly together.

complex mountain see **folded mountain.**

composite volcano a volcano that consists of alternating layers of lava and pyroclastic debris; built up over millions of years, such volcanos are characterized by long periods of dormancy.

compressive stress stress applied to a rock from opposite directions, compressing and flattening the rock mass.

cone of depression the area and shape of a drawdown around a well.

confined aquifer an aquifer overlain by a less permeable bed that keeps the water in the aquifer under pressure.

confining pressure see **geostatic pressure.**

connate water water trapped in the original sediments during deposition and lithification.

contact the plane of separation between any two different kinds of rocks.

contact metamorphic deposit a hydrothermal deposit that results from hot solutions that leave a cooling intrusion and deposit minerals in cracks in country rock.

contact metamorphism the process by which country rock surrounding a hot magma intrusion is metamorphosed by the high heat flow coming from the intrusion.

contamination plume the elongate area of contaminated groundwater that is downgradient from the point source of leakage.

continental divide that topographic ridgeline that separates the streams that flow in opposite directions and empty into different oceans.

continental drift the theory that the continents were once joined together and somehow then split and moved apart.

continental glaciation glaciation that affects a broader, flatter part of a continental land mass than does alpine glaciation.

continental rise a very low-angle ridge of sediment that forms between the lower part of the continental slope and the abyssal plain.

continental shelf a shallow, very gently sloping platform that extends seaward from the edge of a continent.

continental slope an area that extends from the seaward edge of a continental shelf into the deep ocean at an average angle of 4 to 5 degrees.

continent-continent convergence the result when two continents collide.

continuous branch the type of magmatic differentiation in which minerals form continuously during cooling.

contour current a current that flows parallel to the edge of a continental slope.

convection currents currents within a material that are driven mostly by changing temperature gradients.

convergent boundary a fault boundary marked by plates that come together.

core the zone of the earth that includes the inner and outer core.

country rock the surrounding rock that magma invades in the formation of intrusive rocks.

crater the circular depression at the top of a volcano.

craton a continental interior that has been structurally inactive for a prolonged time, usually hundreds of millions of years or longer.

creep a mass-wasting event slow enough that it cannot be detected as it is occurring.

crevasse a deep crack or fissure in a glacier.

cross-bedding a sedimentary structure in which the bedding planes of a particular unit are inclined compared to the bedding of the enclosing rocks.

crude oil a liquid containing hydrocarbons that forms in organic- or fossil-rich sediments and rocks.

crust the outermost zone of the earth, its exterior layer.

crustal rebound the process by which crustal rocks that were downwarped by a glacier's weight slowly return to normal elevation after the glacier's retreat.

daughter products products created as an element undergoes radioactive decay.

debris avalanche a rapidly churning mass of rock debris, soil, water, and air that races down very steep slopes.

debris flow a mass-wasting event in which movement and turbulence occur throughout the mass.

debris slide the rapid movement of a mass of debris as a single unit.

deep sea fan see abyssal fan.

deflation the removal of sediment from a land surface by wind.

delta a thick, roughly wedge-shaped accumulation of sediment deposited at the mouth of a stream.

dendritic drainage pattern a veinlike drainage pattern that develops in a rock type that erodes uniformly, such as granite.

depositional coast a gently sloped coast that has been built up by sediments deposited from longshore drift.

depth of focus of an earthquake, the distance between the epicenter and the focus.

desert an area that receives less than 25 centimeters (10 inches) of rain annually.

desert pavement a large surface of the desert floor that is covered by pebbles and stones that resemble rounded paving stones; caused by deflation or temperature changes.

desiccation crack a crack that develops when a muddy sediment is exposed to air and begins to dry out; these cracks combine to form a polygonal pattern.

detachment fault a low-angle fault above which is often a series of thrust faults and below which is undeformed bedrock.

dewatering the step in lithification in which increasing pressure squeezes out some of the water between sediment particles.

diapir a small magma blob resulting from localized melting of the crust; a component of batholiths.

differential stress stress usually caused by tectonic forces applied to a body of rock from different, but not opposite, directions, stretching the rock mass into an elongate shape.

differential weathering the result of the resistance of some rocks more than other rocks to weathering, creating uneven rates of erosion and sometimes spectacular formations.

differentiation the process by which a magma forms different minerals according to changes in temperature and pressure.

dike an intrusive rock that generally occupies a discordant, or crosscutting, crack or fracture that crosses the trend of layering in the country rock.

dip angle the angle between the horizontal plane and a tilted bedding plane.

dip-slip fault a fault in which movement is parallel to the dip of the fault plane in an up or down direction between the two blocks.

disconformity an erosional contact usually parallel to the bedding planes of the upper and lower rock units.

discontinuous branch the type of magmatic differentiation in which minerals form at discrete temperatures and not continuously during cooling.

disseminated deposit a hydrothermal deposit in which the metal ore is evenly distributed in generally low concentrations throughout large masses of rock.

dissolved load earth material in a stream that has been dissolved into ions and carried in solution.

distributary a small, shifting channel that spreads out across a delta from the main river channel and disperses the sediment load.

divergent boundary a fault boundary marked by plates that move away from one another.

downcutting the erosion directly downward by a stream channel.

drainage basin the area drained by a stream and its tributaries.

drainage divide a ridge that separates one drainage basin from another.

drawdown a local lowering of the water table around a well.

drumlin a long, narrow, rounded ridge of till whose long axes parallel the direction a glacier traveled.

dry wash see **arroyo.**

ductile of a rock, flowing plastically in response to stress.

dust storm a windstorm that carries large amounts of sand or sediment through the air.

earthflow the movement of earth material down a hillside as a viscous fluid; earthflows typically occur on steep slopes with thick soil cover that becomes saturated by heavy rains.

earthquake the ground shaking caused by rocks that suddenly move or jolt in response to tectonic stress.

ebb currents tidal currents preceding low tide.

elastic rebound theory the theory that suggests that in some cases energy is stored in rock that is being bent (deformed) by tectonic forces until the energy in the rock exceeds the rock's chemical bonds and it breaks, releasing the energy and causing motion.

elastic strain strain after which the body of a rock returns to its previous shape when stress has been removed.

end moraine an extensive and typically crescent shaped pile of till built up at the front of a glacier.

ephemeral stream a stream that flows intermittently as a result of periods of sudden rainfall.

epicenter the point on the surface directly above the focus of an earthquake.

erosion the picking up of sediment and soil particles by an agent such as wind or water.

erratic a boulder that has been deposited by a glacier and is not derived from the local bedrock.

esker a long, winding ridge of outwash deposited in streams flowing through ice caves and tunnels at the base of a glacier.

estuary a drowned river mouth from an older coastline; appears as a long arm of ocean water extending inland from the coast.

evaporitic rock a rock formed from minerals that chemically precipitated from water.

exfoliation the process by which curved sheets of rock loosen and fall from a weathered rock surface.

exfoliation dome a large rounded landform (usually composed of intrusive rocks) that results from exfoliation.

exotic terrane a terrane that did not form naturally through accretion and has likely collided with the continental margin.

extrusive igneous rock igneous rock that crystallized from liquid magmas that reached the surface and were generally vented as volcanic lavas.

facies see **metamorphic grades.**

failed rift rifting that ceases before the crustal mass has been separated into parts.

fall a mass-wasting movement in which earth material free-falls from a steep face or cliff and generally collects at the base as talus.

fault an area in which rock has been displaced along a fracture, such as having one side that is moved up or down.

fault block mountain a mountain that is bordered on both sides by steeply dipping faults, such as a horst.

fault gouge the broken material within a fault.

fault plane a plane of fracture in a rock along which movement has occurred.

fault zone a series of parallel fault planes that are close together and form a wider zone of structural weakness.

felsic rock a rock that is rich in silica, potassium, sodium, and aluminum and that contains only small amounts of iron, magnesium, and calcium.

fetch the distance that wind travels over a surface of water.

fiord a steep-walled, fingerlike coastal inlet that was carved by glacial action and later flooded by the rising sea.

firn rounded granules formed by the compaction of snow by pressure from overlying snow and cemented by ice.

first-motion studies studies that indicate whether the first rock motion in an earthquake was a push (the rock moved toward the seismograph station) or a pull (the rock moved away from the station).

fissile splitting naturally along layers.

flash flood flood resulting from very heavy rainfall over short periods.

flood currents tidal currents preceding high tide.

floodplain the area created on both sides of a stream when periodic flooding deposits mud and silt over extensive, low-lying areas.

flow a mass-wasting movement in which the mass moves downslope like a viscous fluid.

flowing artesian well a well that taps an aquifer under confining pressure that is sufficient to force the water to rise naturally to the surface through the well.

focus the point of origin of an earthquake.

fold a bend in a layered rock.

fold and thrust belt a mountain-building event in which rocks are folded during tectonic stress and detached as thin layers along thrust zones, which vertically stack the layers; typically occurs on the continental side of a magmatic arc.

folded mountain a mountain created by intense compressional forces that fold, fault, and metamorphose the rocks, a process that resulted in many of the world's biggest mountain belts.

foliation the alignment of parallel layers or bands of mineral grains in a rock subjected to prolonged differential stress and/or shearing.

footwall the block that underlies an inclined dip-slip fault.

forearc basin the relatively undisturbed expanse of ocean floor between an accretionary wedge and an island arc.

foreland basin a shallow continental basin behind a magmatic arc, a result of subsidence.

foreset bed a sandy bed that composes the main body of a delta.

foreshock one of the small earthquakes that may precede the main earthquake.

fossil the trace of a plant or animal in a sedimentary rock.

fossil assemblage a group of different kinds of fossils that coexisted; more useful than single types of fossils in determining the age of a formation.

fossil fuel coal, oil, or gas derived from organic-rich rocks.

fracture a crack in a rock along which no motion has taken place.

fringing reef a flat expanse of reef that is attached directly to shore.

frost heaving the process by which rock and soil are lifted vertically by the formation of ice and repeated freezing and thawing.

frost wedging the widening and deepening of cracks by ice, breaking off pieces and slabs of rock.

fumerole a vent in or near a volcano from which steam and other gases escape from molten rock below.

gaining stream a stream into which groundwater flows from the saturated zone.

geologic cross section a vertical slice across a map area; depicts the spatial relationships of rock units and structures beneath the surface.

geophysics a field concerning the application of the laws of physics to the study of the earth.

geostatic pressure pressure that is equally applied to all sides of a deeply buried mass of rock.

geothermal energy the energy produced when exceptionally hot water underground is tapped by wells and used to generate electricity.

geothermal gradient the rate at which temperature increases with depth.

geyser an explosive hot spring that periodically erupts scalding water and steam; water temperatures in a geyser are generally near boiling.

geyserite a build-up of ledgelike layers, generally of calcite or silica, around a geyser.

glaciation the movement of an ice mass over a land surface.

glacier a large mass of ice that forms on land during cooler climatic periods.

Gondwanaland a paleocontinent that consisted of what is now Africa, India, South America, Australia, and Antarctica.

gossan a rusty, iron-bearing cap; a remnant of a weathered metallic ore deposit at the surface.

graben a feature formed when a block that is bounded by normal faults slips downward, usually because of a tensional force, creating a valleylike depression.

graded bed a bed in which the base consists of coarser material and subsequent beds grade upward through sand and silt to the finest clay sizes at the top.

graded stream a stream that has smoothed out its longitudinal profile to resemble a smooth, concave-upward curve.

gravity meter a device that measures the force of gravity between a mass inside the instrument and the earth.

groin one of a series of walls built perpendicular to the coast to widen beaches that are losing sand to longshore drift.

ground moraine a thin, widespread layer of till deposited across the surface as an ice sheet melts.

groundwater water derived from rain and melting snow that percolates downward from the surface and collects in the open pore spaces between soil particles or in cracks and fissures in bedrock.

guyot a submerged, flat-topped seamount that was once above sea level and was eroded flat by continual wave action.

half-life the time it takes for half of a known quantity of radioactive material to convert to daughter products.

halo the zone of metamorphism surrounding an intrusion in contact metamorphism.

hanging valley a valley that forms a cliff face with the main valley it enters because its lower part has been eroded away by glacial action.

hanging wall the block that overlies the inclined fault plane in a dip-slip fault.

hardness a quality of minerals determined by the Mohs hardness scale.

hardpan a layer of soil, usually the B horizon, that is so hard (usually cemented by calcite or quartz) that even a backhoe cannot break through it.

headland a rocky arm of a coastline that juts into the sea.

headward erosion erosion that results when a valley is extended upward above its original source by gullying, mass wasting, and sheet erosion.

headwaters the origination of a stream; usually in higher elevations of mountainous terrain.

heat flow of the earth in general, the amount of heat from the earth's interior that is lost at the surface.

heavy crude oil a dense, viscous petroleum that flows so slowly it is usually left behind in an oil field.

herringbone cross-bedding a distinctive pattern of alternating cross-bedding directions that is reflective of a rhythmic, high-energy environment, such as a tidal zone.

hinge line the center axis of a fold.

horn a sharply defined peak that has formed from erosional processes along the rim of a cirque.

horst a feature that results when a block that is bounded by normal faults experiences a compressive force that forces the block upward, forming mountainous terrain.

hot spring a spring with water 6 to 9 degrees centigrade (11 to 16 degrees Fahrenheit) warmer than the mean annual air temperature for the locality where it occurs.

hot springs deposit a disseminated metal deposit formed in response to hot spring activity at the surface of the earth.

hydraulic action the ability of flowing water to dislodge, pick up, and transport rock particles or sediment.

hydrogenous sediments sediments on the ocean floor that have chemically precipitated from seawater.

hydrologic cycle the continuous exchange of water between the atmosphere, oceans, continents, plants, and animals.

hydrothermal deposit of metallic ore, the result of the formation of rich deposits from hydrothermal solutions that circulate through fractured country rock.

hydrothermal rocks those rocks whose minerals crystallized from hot water or whose minerals have been altered by hot water passing through them.

hydrothermal vein minerals that are deposited from hydrothermal processes and fill a crack in the country rock.

iceberg a floating mass of ice that results from calving from a glacial face into the water of a lake or ocean.

ice cap a glacial ice mass that is centered on a highland area and migrates outward in all directions.

ice fall a blocky, piled-up ice surface that results when rapid ice movements rupture glacial crevasses.

ice sheet a glacier that covers a broad expanse of land and is not restricted to a channel.

igneous rock rock that at one time was molten and part of magmas or lavas and that then cooled.

incised meander a steep-walled canyon that results from the downcutting of a meandering stream.

inclinometer a device on a compass used to measure dip angle.

inclusion a rock fragment enclosed within an intrusive rock unit.

index fossil a fossil of those species that lived only during a restricted period and that identifies the narrow time range during which the host rock could have formed.

inner core the spherical, solid, innermost part of the earth.

inselberg an isolated bedrock remnant of a former mountain front; may project through the pediment cover.

interior drainage pattern a drainage pattern in which streams empty into landlocked basins.

intermediate rock a general term for rock between mafic and felsic classifications.

intrusive igneous rock igneous rock that formed from magmas that moved upward into cracks and voids deep in the crust and that never reached the surface.

island arc a curved chain of islands that develops between an oceanic trench and a continental landmass.

isoclinal fold a fold that has undergone stress great enough to compress its limbs tightly together.

isostasy the equilibrium, or balance, between adjacent blocks of crust overlying the mantle.

isotope an atom of an element that contains a different number of neutrons in its nuclei than does another atom of that element.

jetty a wall that is built on both sides of a harbor and that extends into the ocean to protect the harbor from sedimentation and destructive waves.

joint an opening in a rock along which the rock exhibits no displacement; generally an equilibrium response to cooling or unloading.

joint set a series of roughly parallel joints that occur in one direction.

jovian planets those outer planets that have densities of less than 2 g/cm³: Saturn, Jupiter, Uranus, Neptune, and Pluto.

kame a steep-sided mound of stratified till that was deposited by meltwater in depressions or openings in the ice or as short-lived deltas or fans at the mouths of meltwater streams.

karst topography an irregular land surface dotted with numerous sinkholes and depressions related to underlying cave systems.

kettle a depression in glacial till that results when a buried block of glacial ice melts.

kettle lake a body of water occupying a kettle.

komatiite a typical ultramafic extrusive rock that is mostly olivine and pyroxene, with lesser feldspar.

laccolith an intrusive feature similar to a sill but formed from a more viscous magma that creates a lens-shaped mass between sedimentary layers, arching the overlying strata upward.

lahar a mudflow originating on a volcanic slope.

landslide a destructive, rapid mass-wasting event.

lateral erosion erosion that occurs when a stream meanders or braids back and forth across its valley floor or channel, undercutting and eroding its banks.

lateral moraine a moraine consisting of rock debris and sediment that have worked loose from the walls beside a valley glacier and have accumulated in ridges along the sides of the glacier.

laterite a typically bright red, highly leached, residual soil that forms in tropical regions.

lateritic weathering weathering that results in residual deposits that become enriched through the chemical breakdown and removal of more reactive elements of a rock.

Laurasia the paleocontinent that once included the present-day landmasses of North America and Eurasia.

lava magma that is extruded at the earth's surface, as from a volcano.

lava flood nonvolcanic lava that vents from deep cracks in the continental crust.

law of faunal succession a law that states that fossil species succeed one another in undisturbed rocks in a definite and recognizable order around the world.

law of original horizontality a law that states that most sedimentary rocks formed as nearly horizontal layers.

law of superposition a law that states that in an undisturbed sequence of sedimentary rocks or lava flows the overlying rock is younger than the underlying rock.

left-lateral strike-slip fault a strike-slip fault in which the block across the fault appears to have moved to the left.

lignite a soft, brown coal produced by increasing temperature and pressure on peat.

limb one side of a fold.

liquefaction of a landslide, an occurrence in which water-saturated soil moves downslope like a liquid.

lithification the hardening of sediment into a rock.

lithosphere an area composed of the crust and the uppermost part of the mantle.

loam soil that contains about equal amounts of sand, silt, and clay as well as abundant organic matter.

loess silt and clay deposited by wind and weakly cemented by calcite.

longitudinal dune a large, symmetrical ridge of sand that parallels the wind direction; it can be over 100 meters high and over 100 kilometers long.

longshore current a strong current resulting from water being pushed parallel to the shore by repeated wave action; primary transporter of sand in the shoreline environment.

longshore drift the movement of sediment parallel to the shore by wave action.

losing stream a stream whose channel lies above the water table and loses water into the unsaturated zone through which it is flowing.

luster the appearance or quality of light reflected from a mineral's surface.

mafic rock an igneous rock containing approximately 50 percent silica and relatively high percentages of iron, magnesium, and calcium.

magma molten rock that forms below the surface of the earth, usually at depths of 100 kilometers or greater.

magmatic arc a general term for belts of andesitic island arcs or inland andesitic mountain ranges (volcanic arcs) that develop along continental edges.

magmatic deposit of metallic ore, the result when the minerals settle to the bottom of an intrusive body and form thin, high-grade layers.

magmatic water water derived from magmas.

magnetic anomaly an area of magnetism that is either higher or lower than the average magnetic field for that region.

magnetic field of a planet, a magnetic force that surrounds the planet and probably originates from its metallic core.

magnetic pole a locality at which magnetic lines of force converge to create the strongest point in the magnetic field.

magnetometer a device for measuring the intensity of the magnetic field at the earth's surface.

mantle the middle zone of the earth, between the core and the crust.

mantle plume a "hot spot" in the crust where hot mantle material has ascended along deep penetrating cracks in the crust.

marine terrace a broad, gently sloping platform offshore from a beach face.

marker beds those distinctive layers in a sedimentary sequence that allow exposures in different areas to be definitely correlated, or linked.

mass wasting the process of erosion whereby rock, soil, and other earth materials move down a slope because of gravitational forces.

meander a hairpinlike feature of a stream caused by erosion on the outside of a curve and deposition on the inside.

mechanical weathering the process by which rocks are physically broken down into smaller pieces by external conditions, such as the freezing and thawing of water in cracks in the rock.

medial moraine a long ridge of till that results when lateral moraines join as two tributary glaciers merge to form a single glacier.

Mediterranean-Himalayan belt a zone that runs through the Mediterranean region eastward through Asia and to the East Indies marked by frequent earthquake and volcanic activity.

member a distinctive rock layer that is part of a larger rock formation.

mesa a remnant flat-topped tower or column resulting from the weathering and erosion of a plateau's slopes.

metamorphic grades the different groups of minerals that crystallize and are stable at the different pressure and temperature ranges during regional metamorphism.

metamorphic rock a rock created by solid-state transformation (no melting) of a rock mass into a rock of generally the same chemistry but with different textures and minerals.

metasomatism the process by which hot-water solutions carrying ions from an outside source move through a rock mass via fractures or pore space.

meteoric water water that is derived from the atmosphere as rain or snow and that moves down into the bedrock from the earth's surface.

microseism a very small seismic tremor.

midoceanic ridge a deep crustal fault on the ocean floor that separates crustal plates and generates new ocean crust.

mineral a combination of elements that forms an inorganic, naturally occurring crystalline solid of a definite chemical composition.

Mississippi Valley-type deposit a concentration of lead and zinc thought to be deposited in porous limestones and sandstones by low-temperature water that was driven out of deeper sediments by compaction.

modified Mercalli scale a ranking system for the intensity of an earthquake, ranking it from 1 to 12 depending on the amount of resulting damage.

Moho see **Mohorovicic discontinuity.**

Mohorovicic discontinuity the first major boundary of the earth's interior; separates the crust from the underlying mantle.

Mohs hardness scale a scale from 1 to 10 on which the relative hardness of minerals is measured; named after its originator, Friedrich Mohs, a German mineralogist.

moraine an accumulation of till either left behind when a glacier recedes or carried on top of alpine glaciers.

mountain belt long chains of mountain ranges; typically thousands of kilometers long.

mountain-building the process of building mountains through tectonic forces that deform, metamorphose, and uplift crustal rocks.

mountain range a group of mountain peaks or ridges that form a discrete topographic area.

mountain root a bulge of continental crust downward into the mantle beneath a mountain.

mouth the terminus of a stream.

mud crack see **desiccation crack.**

mud flow the movement of a liquidy mass of soil, rock debris, and water down a well-defined channel.

mudpot a kind of hot spring that produces boiling mud and releases sulfurous gases.

natural attenuation the natural breakdown of groundwater contaminants over time and distance from the point source.

natural gas a gaseous mixture of hydrocarbons that usually occurs with crude oil.

natural levee a ridge of sand and silt deposited near the edge of a stream's channel.

neap tides the lowest tides, which occur close to the first and third quarters of the moon.

nebular hypothesis the hypothesis that suggests that the planets and moons in the solar system formed from a huge hydrogen-helium cloud.

negative budget of a glacier, the losing of more volume than that gained from new snowfall.

negative gravity anomaly the gravity reading of a rock if it is lower than the normal regional gravity value.

negative magnetic anomaly a magnetic reading that is lower than the average regional magnetic field strength.

negative polarity of a rock, the polarity created when the earth's magnetic field was reversed, which reduced the earth's net field strength.

nonconformity the erosional contact that separates a younger sedimentary rock unit from a plutonic or metamorphic rock unit.

nonflowing artesian well a well in which water from the tapped aquifer must be pumped to the surface.

normal dip-slip fault a dip-slip fault in which the hanging wall block has moved downward relative to the footwall block.

normal force force that is parallel to the surface of the slope.

normal polarity see **positive polarity**.

nose of a fold, its tip.

obduction the process by which a crustal plate overrides another plate for a short distance.

oblique-slip fault a fault in which the fault blocks show both horizontal and vertical displacement.

ocean-continent convergence convergence that occurs when oceanic crust is subducted under continental crust.

oceanic trench a narrow deep trough that parallels the edge of a continent, island arc, or convergence of two oceanic plates and forms at the edge of a subduction zone.

ocean-ocean convergence convergence that occurs when two plates carrying ocean crust converge, with one slab subducted under the other at an ocean trench.

oil field the occurrence of multiple oil pools in one area.

oil sands sandstone deposits that have been cemented with tar or asphalt.

oil shales organic-rich shale formations from which oil can be extracted.

open fold a broad feature in which the limbs of a fold dip at a gentle angle away from the crest of the fold.

ophiolite a mafic rock sequence at the earth's surface that is believed to be pieces of ancient oceanic crust that were thrust onto the continent during subduction and mountain-building.

organic sedimentary rock a sedimentary rock composed primarily of accumulations of organic remains from plants or animals.

orogenesis see **orogeny**.

orogeny the folding, faulting, deformation, and metamorphism from the onset of intense tectonic stress that results in mountain-building.

outcrop a bedrock exposure at the surface of the earth.

outer core the outer shell of the core, between the mantle and the inner core, that is inferred to be molten (liquid).

outwash the sediments deposited by glacial meltwater.

outwash plain the broad front of outwash associated with an ice sheet.

overturned fold a fold whose limbs dip in the same direction, indicating that the upper part of the fold has overridden the lower part.

oxbow lake a body of water shaped roughly like a U and formed when a meander begins to close on itself and the stream breaks through and bypasses the meander.

pahoehoe flow a basalt flow with a ropy or undulating surface resulting from quick cooling and solidification of the lava.

paleocoast an old coastline that has been preserved in the geologic record.

paleocurrent a direction of sediment transport or ice movement that is revealed by sedimentary or glacial features.

paleomagnetic field an ancient magnetic field that can be detected from the orientation of magnetic crystals in rocks such as basalt.

paleontology the study of fossils.

Pangaea a single continental mass that rifted to form our present-day continents.

parabolic dune a deeply curved dune with the tips pointing into the wind; usually forms around a blowout in vegetated areas.

parallel retreat the retention by slopes of their original steepness as they erode.

parent rock the original rock from which a metamorphic rock was formed.

partial melting the process by which a portion of the magma that is forming from a melting mass of rock separates and rises as a distinct magma.

passive continental margin marked by a landward, continental shelf followed by a deeper continental slope, continental rise, and flat abyssal plain; characterized by a lack of earthquake activity.

P body wave a compressional (longitudinal) body wave that induces rock to vibrate parallel to the direction the wave is traveling.

peat an unlithified organic material that can be cut into blocks and burned for fuel.

pedalfer a thick soil high in aluminum and iron that develops in response to abundant rainfall, organic acids, and strong downward leaching.

pediment a low-angle erosion surface at the foot of a mountain range that is typically covered by up to 100 feet of sediment; occurs between the bajada and the range front.

pedocal a thin, poorly leached soil formed in arid climates by the upward movement of soil water by subsurface evaporation and capillary action.

pegmatite a dike that contains very coarse-grained crystals.

pelagic sediment a sea-floor sediment that is composed of fine-grained clay particles and microskeletons of marine organisms that settle slowly to the ocean floor; its clay component (and sometimes volcanic ash) is generally carried from land by wind and deposited on the surface of the ocean.

peneplain an area reduced by erosion nearly to a plain.

perched water table an accumulation of groundwater that is held above the water table in the unsaturated zone by an impermeable bed such as clay.

permeability of a rock, the ease with which fluid is transmitted through its pore space.

petroleum a general term that includes both natural gas and crude oil.

physical geology the study of the earth's rocks, minerals, and soils and how they have formed through time.

piedmont glacier the forwardmost extension of a valley glacier; forms where the ice emerges at the front of the mountain range.

pillow structures blobs of submarine lava that break through the thin, hardened exterior of a lava flow and chill immediately in the cold water, forming small rounded shapes.

placer deposit a deposit of heavy metallic minerals, such as iron or titanium minerals, or native gold or diamonds, that have been concentrated by wave or water action in a river or beach environment.

plan geologic map a two-dimensional map showing the locations and shapes of the outcrops at an appropriate scale and indicating, through a variety of geologic symbols, features such as folds, faults, contacts between different rock units, and strike and dip.

plastic deformation the physical, permanent changes, such as folds or stretching, in a rock from tectonic forces that do not result in fracturing.

plastic flow the ability of a material, such as glacial ice, to flow plastically without breaking.

plastic strain strain that results in a permanent change in the shape of a rock.

plate a segment of the earth's crust that is bounded by deep faults and moves in response to internal forces.

plateau a flat-lying hill underlain by resistant rock.

plateau basalt see **lava flood.**

plate tectonics the theory that the earth's surface is divided into large, slow-moving crustal plates that are driven by internal forces, such as convection currents in the mantle.

playa lake a lake formed from water that drains from mountains into the central part of a valley.

plunge the angle between the horizontal and the hinge line in a plunging fold.

plunging fold a fold that has been tipped by tectonic forces and that has a hinge line, or axis, that is not horizontal.

pluton see **plutonic rock.**

plutonic rock an intrusive, discordant, generally coarse-grained rock that was formed deep in the earth's crust.

pluvial lake a lake formed during the wetter climates that existed during and after glacial retreat.

point source the point of contamination.

polar wandering the apparent movement of the earth's geographic and magnetic poles through geologic time.

pore space open space between sediment grains.

porosity of a rock or sedimentary deposit, the percentage of volume that consists of voids and open space.

porphyritic of an igneous rock, containing coarser crystals that are supported in a fine-grained groundmass.

porphyry copper deposit a disseminated deposit in which copper and molybdenum are found in porphyritic intrusive rocks.

positive budget of a glacier, the gaining of more volume from new snowfall than the losing from melting.

positive gravity anomaly the gravity reading of a rock if it is higher than the normal regional gravity value.

positive magnetic anomaly a magnetic reading that exceeds the average magnetic field strength.

positive polarity of a rock, when its magnetic field is the same as the earth's field today.

pothole a circular depression eroded into the bedrock of a stream by abrasive sediments.

primary body wave see **P body wave.**

protolith see **parent rock.**

provenance area the source from which sediment originated.

P-wave shadow zone that area on the earth's surface in which P waves from an earthquake cannot be detected.

pyroclastic cone see **cinder cone.**

pyroclastic debris fragments of rock ejected from a volcano.

pyroclastic flow a dense mixture of hot gas and pyroclastic debris.

radial drainage pattern a drainage pattern that resembles the spokes on a wheel; occurs when the streams originate on the flanks of conical mountains.

radioactive decay the spontaneous breakdown of isotopes that contain unstable nuclei.

rain shadow an area on the lee side of a mountain range that is arid because most of the rain is precipitated on the other side of the range.

receding glacier a glacier that, although it can move downslope, cannot overtake its rate of uphill recession.

recessional moraine a moraine that develops at the front of a receding glacier.

recharge the process by which new water is added to the saturated zone, replenishing water that is lost.

rectangular drainage pattern a drainage pattern created in bedrock that is regularly fractured or jointed in 90-degree angles.

recumbent fold a fold so overturned that its limbs are essentially horizontal and parallel.

reef an accumulation of organisms (typically corals and algae) that forms in warm, shallow ocean environments; resistant ridge that rims islands, lagoons, and other shorelines.

regional metamorphism metamorphism of rocks typically exposed to tectonic forces and associated high pressures and temperatures.

regolith the interface between bedrock and overlying sedimentary material; consists of solid fragments of weathered rock.

renewable resource a resource or commodity that can be replenished, such as trees and crops.

reserves that subgroup of a resource that has been discovered and can be extracted at a profit.

reservoir rock a rock with the required permeability and porosity to hold large accumulations of petroleum.

residual soil soil developed from the weathering of the underlying bedrock.

resource that amount of a geologic commodity that exists in both discovered and undiscovered deposits.

reverse dip-slip fault a dip-slip fault in which the hanging wall block has moved upward relative to the footwall block.

Richter scale a numerical scale that lists earthquake magnitude in logarithmic increments from about 2 to 8.6.

ridge-push a term that refers to the cooling and sinking of new crust as it moves away from a midoceanic ridge along a deeper lithospheric plane of weakness.

rift valley a large crack in the crest of a midoceanic ridge that typically forms a graben-type valley.

right-lateral strike-slip fault a strike-slip fault in which the block across the fault appears to have moved to the right.

rill a concentration of sheetwash into a small channel; rills merge to form larger streams.

Rim of Fire see **circum-Pacific belt.**

rip current a narrow channel of water that flows straight back out to sea after its waves have broken on the beach.

ripple marks gentle, repeated ridges, usually in sand or silt, that form perpendicular to the flow of wind or water.

rock a solid aggregate of bonded mineral crystals.

rock avalanche the rapid descent of a mass of variously sized rock fragments.

rock-basin lake a depression that is scoured out by an advancing glacier and later fills with water.

rock cycle the various interrelated ways rock types form from geological processes.

rock formation an occurrence of rock with a set of characteristics that distinguishes it from the rocks above or below it.

rock slide the rapid movement of loose rock along an inclined plane.

rounding the smoothing of rock fragments during transportation.

saltation of water, the process in which turbulent or eddying currents temporarily lift larger sediment grains into the overlying flow of water.

saltation of wind, the process in which air currents temporarily lift larger sediment grains into the air.

salt dome a vertical column of rock salt that extends upward through a sedimentary sequence, creating folds and faults that trap petroleum.

salt flat a flat surface area covered by salt that precipitated by evaporation.

sand dune a heap of loose sand deposited by wind action.

sand fall a mass of sand that dislodges and falls in a submarine canyon.

saturated (saturation) zone rock and soil in which all the porosity is filled with water.

S body wave a body wave, only about half as fast as a P wave, that causes rock to vibrate perpendicularly to the direction of wave travel.

scarp a steep hillside or cliff that typically results from faulting or mass wasting.

sea arch a stack whose center has been eroded through, producing a bridgelike shape, because the rock is softer or more fractured.

sea cave a cavity eroded into a sea cliff in the wave zone.

sea cliff a steep slope along a coastline that results from the slope's base being eroded by waves.

sea floor spreading the process by which new basaltic oceanic crust forms at a midoceanic ridge and is slowly pushed away on both sides toward the continents as more new crust is produced.

seamount a conical, usually basaltic volcanic mountain that forms on the ocean floor.

secondary body wave see **S body wave.**

sedimentary rock a rock that is composed of sediment grains that have been compacted and lithified.

sedimentary structures features that were part of sediments when they were deposited and which were preserved when the sediments became lithified.

seif see **longitudinal dune.**

seismic gap a stretch along an active fault zone that has not produced earthquakes for a significant time.

seismic reflection the return of some of the energy from seismic waves that have penetrated downward from the surface or near-surface, hit a rock boundary, and bounded back to the surface.

seismic refraction a change in the direction of travel of a seismic wave as it passes through different mediums; occurs only if the mediums have different densities or strengths, which change the velocity of the seismic wave.

seismic sea wave see **tidal wave.**

seismic wave a wave of energy that is released by an earthquake.

seismogram the series of squiggly lines recorded by a seismograph.

seismograph a device used to record the motion of a seismometer during an earthquake.

seismometer a suspended pendulumlike device used to detect seismic waves.

shear force force that is parallel to the surface of the slope.

shearing the sliding motion that is parallel to and results from compressive forces applied to a rock mass.

shear plane the surface along which shearing occurs.

shear strength an object's resistance to movement that needs to be overcome in order to make it move.

shear stress stress that results when forces from opposite directions create a shear plane in an area in which the forces run parallel to one another.

sheet joint a crack that parallels the outer surface of a rock.

sheetwash a thin layer of unchanneled water that flows downhill during very heavy rains.

shield volcano a broad, cone-shaped hill or mountain made from solidified lava flows.

silica tetrahedron four oxygen atoms connected to a smaller, central silicon atom.

sill an intrusive body formed from magma that entered country rock parallel to the bedding and is thus concordant with the country rock.

sinkhole a basinlike depression at the surface caused when a portion of a cave system collapses.

sinter a build-up of ledgelike layers, generally of calcite or silica, around a hot spring.

slab-pull a term that refers to the result of the cold edge of a plate subducting at a steep angle through the mantle, its downward motion tending to pull the plate away from the ridge crest.

slide a mass-wasting movement that moves along a surface parallel to the slope of the surface.

slip a mass-wasting movement in which the mass moves as a single unit along a well-defined surface or plane.

slip face the steeper, downwind slope of a sand dune.

slot canyon a vertical-walled canyon where mass-wasting processes have been very limited.

slump a mass-wasting movement along a curved surface where the downward movement of the upper part of the mass leaves a steep scarp and the bottom part is pushed outward along a more horizontal plane.

snow line of a glacier, the irregular boundary between the zone of accumulation and the zone of wastage.

soil layers of weathered, unconsolidated particles of earth material that contain organic material and can support vegetation.

soil horizon one of the three layers of mature soil.

solifluction a variety of earthflow in which the flow of water-saturated earth is over an impermeable surface such as permafrost; usually occurs in bitterly cold regions.

solum the O, A, and B soil horizons.

solution weathering the process by which certain minerals are completely dissolved by acidic solutions.

sorting the process by which large, coarse, angular pieces of sediment are deposited near a source area, while progressively smaller and smoother sediments are carried farther.

spatter cone a smaller feature usually associated with an already extruded and cooling lava flow from a shield volcano.

spherical weathering weathering that occurs when the corners of an angular rock are broken down more quickly than the flat surfaces, forming rounded shapes.

spit a fingerlike ridge of sand that projects into a bay.

spreading axis see **spreading center.**

spreading center a divergent boundary (midoceanic ridge) along which new oceanic crust is formed and pushed outward.

spring tides tides that occur at the times of the new and full moons; spring tides exhibit the greatest difference in tidal elevations.

stability field the ranges of temperature and pressure in which a particular mineral is stable.

stack an erosional remnant of a sea cliff; a stack is rooted to the wave-cut platform and stands above the surface of the water.

star dune an isolated hill of sand formed by variable winds in the Sahara and Arabian deserts; the base of the dune resembles a multipointed star.

stock a pluton that occupies less than 100 square kilometers at the earth's surface.

strain a change in the volume and/or shape of a rock because of stress.

stratigraphic trap a naturally occurring change in a sedimentary sequence that traps migrating oil and gas; examples include a lens of sandstone in a larger bed of shale or a porous reef structure in a limestone unit.

stream base level the elevation of a stream's most horizontal flow and lowest velocity.

stream capacity the total load of sediment a stream is capable of carrying.

stream competence a measure of the largest-sized particle a stream can transport.

stream discharge the volume of water that flows past a certain point in a certain amount of time.

stream gradient the downhill slope of a channel; typically measured in feet per mile.

stream terrace a steplike bench that occurs above a stream bed and floodplain and that is cut into bedrock or is a remnant of older river sediments that have since been eroded.

stream valley a topographically low area, typically centered on a stream, that is produced by mass wasting and erosion.

stream velocity the speed at which a stream flows.

stress an applied force (usually tectonic) that tends to physically alter a rock mass.

strike the compass bearing of the line formed by the intersection of a tilted bedding plane with the horizontal plane.

strike-slip fault a fault in which the blocks on either side of the fault move horizontally in relation to each other, parallel to the strike of the fault.

structural basin a variation of a syncline in which all the beds dip inward toward the center of the basin.

structural dome a variety of anticline, a feature of which is that the central area has been warped and uplifted and all the surrounding rock units dip away from the center.

structural geology the study of the processes that result in the formation of geologic structures.

structural trap a structure such as a fault between reservoir rocks and impermeable rocks, a thrust fault, or a fold such as an anticline that traps migrating petroleum.

subbituminous coal weakly metamorphosed, black, soft, sooty coal.

subduction the process by which oceanic crust is pushed against, and finally underneath, continental or oceanic crust.

subduction boundary a convergent boundary marked by the oceanic crust of one plate that is being pushed downward beneath the continental or oceanic crust of another plate.

subduction complex see **accretionary wedge**.

subduction zone the gently dipping zone along which subduction occurs.

submarine canyon a V-shaped erosion feature that cuts a continental shelf and slope.

subsoil the layer of soil that underlies the topsoil.

supergene deposit a high-grade metal deposit enriched through the processes of weathering.

surf the zone where waves break against a shoreline.

surface wave the slowest of the seismic waves; surface waves travel outward on the earth's surface from the epicenter much as ripples do from a stone thrown into the water.

suspect terrane a terrane that does not fit the regional pattern or has conflicting age dates.

suspended load of a stream, the fine-grained sediment that remains in the water in a stream during transportation.

suspended load of wind, the fine-grained clay and silt that is carried long distances.

suture zone the line of collision at a convergent boundary, typically continent-to-continent.

swash the still-turbulent sheet of water that sweeps up the slope of a beach.

S-wave shadow zone the area on the earth's surface in which S waves from an earthquake cannot be detected.

syncline a fold that arches downward to form a trough.

talus the accumulation of rock debris at the base of a steep slope.

tarn see **rock-basin lake.**

tar sands see **oil sands.**

tectonostratigraphic terrane see **terrane.**

tensional stress stress that occurs when a rock is subjected to forces that tend to elongate it or pull it apart.

tephra pyroclastic debris that is ejected from a volcano.

terminal moraine a ridge of till that marks the farthest advance of a glacier before it started to recede.

terminus of a glacier, the front.

terrane a region of geologic continuity distinct from neighboring regions.

terrestrial planets those that have densities of $3g/cm^3$ or more: Mercury, Venus, Earth, and Mars.

terrigenous sediment a sea-floor sediment derived from land and usually deposited on the continental shelf, continental rise, and abyssal plain.

texture a term describing the sizes and orientations of a rock's mineral or rock fragment components.

theory of glacial ages a theory proposed by Swiss naturalist Louis Agassiz that parts of the earth's surface in the geologic past were covered with larger glaciers than we see today.

thermal metamorphism see **contact metamorphism.**

thrust fault a reverse fault in which the hanging block (upper plate) has overridden the footwall block (lower plate) at a very shallow angle for an extensive distance.

tidal current the horizontal flow of water that accompanies the changing tide and flows in two opposite directions.

tidal delta sediments deposited by the back-and-forth tidal action between barrier islands.

tidal flat a flat, muddy zone of coastline affected by tidal currents.

tidal wave a gigantic wall of water, sometimes as high as 90 meters, caused by a submarine earthquake.

tide the rhythmic rise and fall of sea level along a coastline.

till the unsorted and unlayered rock debris and sediment that is carried or later deposited by a glacier.

tombolo a bar of sediment that connects an island to the mainland, forming a small peninsula.

topset bed a nearly horizontal layer of sediment deposited by distributaries as they flow toward a delta front.

topsoil the upper part of a section of loam; topsoil has the highest organic content of types of soil and is considered to be the most fertile.

transform boundary a fault boundary marked by plates that slide past one another.

transported soil soil deposited by agents such as ice and water and not derived from the underlying bedrock.

trap a stratigraphic or structural feature of high porosity that traps migrating petroleum.

travel-time curve a plot of the arrival times of seismic waves relative to distance.

trellis drainage pattern a drainage pattern consisting of a main stream with short tributaries on either side; forms in areas of tilted sedimentary rocks that create parallel ridges and valleys.

trench-suction a term that refers to the subduction of a plate at a steep angle, which pulls the overlying plate and the trench toward the midoceanic ridge.

triple junction (point) the junction of three major faults, thought to be in response to an underlying mantle plume, that signals the onset of rifting.

truncated spurs topographic spurs along a valley that have been truncated by glacial erosion in the valley.

tsunami see **tidal wave.**

tuff a volcanic rock consisting of small particles such as ash and dust.

tuff breccia a volcanic rock that contains angular, coarse rock fragments in a matrix of finer-grained ash and dust.

turbidites sediments that are deposited by turbidity currents and that typically show graded bedding.

turbidity current a large volume of dense, sediment-laden water that results when sand and mud on a continental slope are dislodged by landslides or earthquakes and become suspended in the water.

turbidity flow see **turbidity current.**

ultramafic rock rock consisting almost entirely of ferromagnesian minerals and having no feldspars or quartz.

unconfined aquifer an aquifer that does not have a confining bed that separates the zone of saturation from the unsaturated units above it.

unconformity an erosional contact between two rocks in which the upper unit is usually much younger than the lower unit.

ungraded stream a stream that is still actively downcutting its course and smoothing out irregularities in its gradient through erosion.

uniformitarianism the principle that the geologic processes we see today were active in the geologic past.

unloading the removal of the overlying weight and pressure through erosion when a rock mass is uplifted to the surface, resulting in the mass's slow expansion.

unsaturated zone rock and soil in which pore spaces contain both air and water and therefore are not saturated.

uplifted coast a former coast that has been lifted above the present coastline by tectonic activity.

uplifted marine terrace a former marine terrace that has been lifted above the present coastline by tectonic activity.

upwarped mountain a mountain that is the result of broad arching of the crust or great vertical displacement along a high-angle fault.

valley glacier a mass of ice restricted to high mountain valleys.

valley train the outwash plain of an alpine glacier.

varve one light-colored bed and one dark-colored bed of sediment that form at the bottom of a glacial lake and that represent a single year's deposition.

ventifact a rock that has flattened surfaces formed by windblown sand.

viscosity resistance to flow; a lava with low viscosity spreads quickly, and one with high viscosity flows sluggishly.

volcanic arc a range of andesitic volcanic mountains that forms on the continental edge above a subduction zone.

volcanic dome a rounded volcanic feature, formed from thick, viscous magma, that creates a plug in the vent of a volcano.

volcanic mountain the result of the accumulation of a large amount of volcanic lavas and pyroclastic material around a volcanic vent.

volcanic neck a rock that formed in the vent or throat of a volcano at the end of its eruptive life and remains standing after the flanks of the volcano have eroded away.

volcanism the venting of liquid magma at the surface of the earth.

volcano a hill or mountain that forms around a volcanic vent and that consists of cooled lava, rock fragments, and dust from the eruptions.

water table the contact between the saturated and unsaturated zones.

wave crest the top of a wave.

wave cut platform a flat-lying bench of eroded rock left behind by a sea cliff's retreat.

wave height the vertical distance between the top of the wave and the low point of the wave.

wavelength the horizontal distance between two crests or two troughs of adjacent waves.

wave refraction a process by which breaking waves become more parallel with the shore.

waves of oscillation waves in the open sea; so named because of the orbital motion of water particles in them.

waves of translation waves that begin to break as they meet the shore.

wave trough the low point of a wave.

weathering the breaking apart of rock at the surface through chemical and physical processes.

xenolith a fragment of country rock torn away during the emplacement of magma; generally most abundant near the contact with the country rock.

zone of accumulation of a glacier, the higher portion that is perennially covered with snow.

zone of accumulation of soil, see **B horizon.**

zone of fracture the more rigid portion of a glacier near its surface.

zone of wastage of a glacier, the lower portion, where the ice is lost.